To The Instructor

When two or more teachers of health-oriented chemistry get together, their discussion invariably centers on the problems of presenting a laboratory sequence that (1) stresses the fundamentals, (2) is interesting, relevant, and not too difficult mathematically, and (3) does not require overly expensive instrumentation or supplies. The array of experiments in this manual are the result of many years of experience with these problems.

This LABORATORY MANUAL is designed to accompany the text, *Elements of General, Organic, and Biological Chemistry,* Ninth Edition, by John R. Holum. It is also appropriate for any one-semester course treating a survey of chemistry at this level. The experiments have been used by students and have been frequently revised following student polls regarding clarity and interest and suggestions from instructors. The questions on the Report and Observation Sheets have again been adjusted in the light of student comment—invited or overheard—such as, "What on earth is this one driving at?" And more room for answers has been provided on many Report Sheets.

The section on Important Techniques and Common Laboratory Operations, much referred to in the directions for experiments, contains, in addition to descriptions of techniques, common-sense safety procedures, discussion of significant figures, and suggestions for organizing data and writing useful reports.

A new feature of this Ninth Edition is that we have highlighted any and all instructions for laboratory procedures in a different type face, as well as to print the word "Procedure" in boldface. This avoids the problem of having students neglect to do something because the instructions were lost within the discussion and fragmented from the rest of the instructions. It will look like this:

Procedure. Below is a list of substances to be placed in separate test tubes and then mixed. Note the appearance of each substance before, and after mixing. In each case determine the temperature of the solution(s) before and then immediately after mixing. Apply the criteria of chemical change and physical change to each situation; decide whether there was any change at all.

Also, special instructions that pertain to safety are set off in italics, for example:

Warning: Do Not Smell Or Taste Any Substance On Display!

Repetition and review are methods for presenting facts and relations as well as techniques in such a way as to make them a permanent part of a student's training. This is encouraged by continual reference to experiments completed earlier, and repetition of techniques in different contexts. In some experiments, the students are encouraged to bring samples of personal interest for testing by a specific procedure. The experiments identifying unknowns — a carbohydrate, a protein, and a lipid — seem to develop in students a feeling of confidence through repetition and reliance on their own observations, wherein they must recognize the evidence of tests, as well as organize their laboratory approach and their written reports.

The use of hazardous materials, such as mercury, benzidine, benzene, chloroform, and carbon tetrachloride, is avoided throughout. The use of dichloromethane (methylene chloride) has also been reduced. Hexane is substituted wherever practical. Cautions and warnings are expressed in the directions for any procedure that could have hazardous complications. The findings of OSHA and NIOSH have been taken into account. While it is thought by the authors that the disposal of hazardous waste should follow the dictates of local health authorities, some further suggestions are occasionally given in the TEACHER'S MANUAL. Two excellent publications, available from the American Chemical Society, are "Safety in Academic Chemistry Laboratories," and "Health and Safety Guidelines for Chemistry Teachers," and a thorough review of safety problems was published in Science, 211 (Feb. 20) 777-80 (1981): "Prudent Practices for Handling Hazardous Chemicals in Laboratories" by Blaine C. McKusick.

This MANUAL contains many more experiments than could possibly be finished even in a one-year course, providing the instructor with much individual choice in the laboratory work on fundamentals, as well as a choice of experiments that are of interest to students. The large variety of experiments in the various sections should solve the problem of "lab getting ahead of lecture;" by careful planning, the laboratory work can reinforce the current lecture topic. Experiments in the area of metabolism can become a problem in the designing of the lab at this level; attempts to meet this use models of metabolic pathways.

Special thanks for advice and encouragement go to the entire staff of the Western Connecticut State College, with special gratitude for Dr. Robert J. Merrer's help with the problems of toxicity and Dr. Catherine Hickey-Williams' suggestions for improvement in various experiments. Also, special thanks to Dr. Arlin Gyberg of Augsburg College for his suggestions and comments.

Accompanying this LABORATORY MANUAL, and the textbook *Elements of General, Organic, and Biological Chemistry*, Ninth Edition, is a TEACHER'S MANUAL, with many features designed to save instructors considerable time, available without charge from John Wiley and Sons, 605 Third Ave., New York, New York, 10158. It covers the experiments individually, with lists of supplies and equipment, and frequent suggestions not covered herein,

LABORATORY MANUAL

NINTH EDITION

ELEMENTS OF GENERAL, ORGANIC, AND BIOLOGICAL CHEMISTRY

JOHN R. HOLLUM
Augsburg College

SANDRA L. OLMSTED
Augsburg College

JOHN WILEY & SONS, INC.
New York • Chichester • Brisbane • Toronto • Singapore

ISBN 0-471-05886-6

Printed in the United States of America

10 9 8 7 6 5 4 3

concerning the conducting of an experiment. It has an alphabetical listing of all solutions needed, with full directions for their preparation, cross-referenced by Experiment number. Included are a list of supplementary supplies, and a chemicals purchase list. A special feature is the reproduction of the Report Sheets from the experiments with answers written in.

To The Student

Why a *laboratory* course? Why go to all the trouble of spending two or three hours a week in the laboratory working at things that some of you may feel you "don't understand," and then having to clean up afterwards!?

Actually, it is the very purpose of laboratory work that you *do understand*. The experiments in this book were written to help you see how the main ideas in each experiment relate to the chemical principles you are studying in your textbook, and *why* you should do things in the order prescribed and with the precautions listed.

We hope that a new feature of this Ninth Edition will make the instructions for a laboratory procedure more clear than in previous editions. We have highlighted any and all instructions for laboratory procedures in a different type face, as well as to print the word "Procedure" in boldface. This helps you to find "what you are supposed to do," and avoids the problem of having you neglect to do something because the instructions were lost within the discussion and fragmented from the rest of the procedure. It will look like this:

Procedure. Below is a list of substances to be placed in separate test tubes and then mixed. Note the appearance of each substance before, and after mixing. In each case determine the temperature of the solution(s) before and then immediately after mixing. Apply the criteria of chemical change and physical change to each situation; decide whether there was any change at all.

Also, special instructions that pertain to safety are set off in italics. Please pay particular attention to these, for example:

Warning: Do Not Smell Or Taste Any Substance On Display!

There are many more facets to laboratory work than reinforcing and amplifying text and lecture material. Not the least important of these is to increase your self-confidence by showing you that with a little practice you can handle measuring instruments, make dilutions, make calculations, and do many laboratory manipulations quickly, easily, and effectively. Laboratory work also helps to increase your ability to make significant observations, and to think in relative terms. At the same time, you will find that you have become better able to organize procedures and work efficiently. Further, many of the potential leaders, teachers, supervisors, and scientists who take this course will be given the opportunity to develop their abilities to record and correlate data in a logical manner, so that others can read their reports and understand the results and significance of the tests.

Moreover, there are phases of laboratory work that are particularly important to people who will be checking, handling or administering dosages.

A dose can be totally useless if too small, or fatal if too large. Sometimes the margin for error is very narrow. Therefore, we add to the purposes of the laboratory experience these three: **instilling care in reading labels, in reading directions and following them, and finally in learning the great importance of weighing and measuring accurately.** Here are a few suggestions to make your laboratory experience more fruifful.

1. If an experiment refers you to a technique or a precaution described in another part of the manual, flip over to that page and do a quick review.

2. Make complete records as you proceed. Be sure your numbers are labeled. Don't try to remember a mass, a color, or a sequence of events in a test. *Write it down immediately, on the Report Sheet, and not just on a scrap of paper.*

3. Go over all your Report Sheets when you get them back, make any corrections designated, and keep the Report Sheets in a binder for *review before examinations.*

4. If you do not understand something or seem to be getting an anomalous result, it is important that you consult your instructor. Few people can perform an experiment and get perfect results the first time. On the other hand, we are not trying to achieve the perfection necessary for clinical analysis. The instructor is in the laboratory to help you get the most out of your experiment the first time around. Don't cheat yourself out of this opportunity—*make sure you do understand!*

Contents

LABORATORY MANUAL

NINTH EDITION

ELEMENTS OF GENERAL, ORGANIC, AND BIOLOGICAL CHEMISTRY

CHAPTER 1

Laboratory Operations, Reporting Data, and Safety

How to Protect Yourself in the Laboratory:
SAFETY RULES

Every experiment in the Manual is designed to minimize hazards, but the following rules are a necessary adjunct to that design.

1. Wear safety glasses at all times when you are in the laboratory. The type to be worn is often a matter of state or municipal law. Your instructor will advise you. Those who wear prescription glasses have considerable protection already. At the very least, others should wear inexpensive plastic nonprescription glasses.

By "safety glasses" we mean industrial-quality eye protective devices meeting the standards of the American Standard Safety Code for Head, Eye, and Respiratory Protection, presently defined as ASA Z2.1-1959.

Wearers of **contact lenses** should be aware of the serious problems which can result from irritating vapors or liquids which can get under the lens or be absorbed by it (especially in soft lenses), increasing the exposure of the eye to the trauma. This problem is thoroughly discussed in *Chemical and Engineering News*, Nov. 19, 1979, p. 3, 4, and 84, and in the same journal, April 7, 1980, pages 2 and 65. It is evident that contact-lens wearers should wear the same protection as those who do not wear prescription glasses, **and should tell the instructors that they wear contact lenses.**

2. Learn the exact locations of eyewash fountains, fire extinguishers, fire alarms, fire blankets, and other safety features in your laboratory, as well as how to use these devices. Sketch the laboratory and indicate their locations.

3. Work only during the scheduled laboratory periods and perform only authorized experiments. In many smaller schools, laboratories are open for students to use at any time during the day when another laboratory section is not using them. Your instructor will advise you about local regulations. An important safety rule, however, is *never work alone in a laboratory*. If an accident occurs, the other person may be able to aid you. If you need to go to the health service, have another person accompany you.

4. Wash your hands thoroughly when you leave the laboratory for any reason, and when you finish your work. **Neither food nor beverage is allowed in the laboratory.**

5. If you feel faint, sit down right away.

6. If you are burned and require the attention of a doctor, have someone accompany you to the doctor's office. Do not apply salves or ointments on the burned areas; let the doctor decide what treatment is needed. *Prompt cooling of a burned area with cold water markedly reduces subsequent pain and facilitates healing of the area.*

7. Read labels. Some accidents happen when labels of bottles are not read carefully. Get in the habit of reading out loud (but softly) the label of a bottle you intend to take from the shelf. You will be more conscious of what you are doing.

8. Experiments go awry and accidents may result when chemicals become contaminated. To **avoid contamination** (a) discard unused chemicals; do not return them to reagent bottles; clean up anything you spill; never put a medicine dropper or a pipet from your desk into a reagent bottle. If a dropper is not supplied, pour a very small amount of the reagent into one of your clean, dry beakers or flasks and use this supply to transfer dropper or pipet quantities; try to keep inner walls of bottle stoppers or corks from touching tops of desks or shelves where they might pick up dust or other chemicals. If a stopper has a flat top, it may be rested upside down on the shelf or bench. If it has a penny head top, the stopper may be held between the fingers as shown in Fig.1, or place it, or other types of stoppers, on the surface furnished for them—not on the lab table. (Watch glasses, petri dishes, or plastic lids such as those from coffee cans make excellent bottle-top holders.)

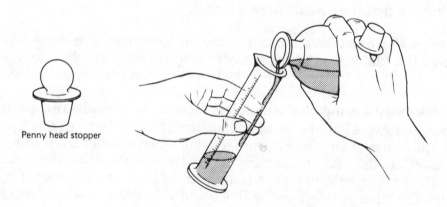

Penny head stopper

Fig. 1. Dispensing liquids. A penny head stopper may be held as shown while the solution is poured.

9. Discard all waste solids—water-insoluble chemicals, litmus paper, filter paper, used matches, broken glass, paper towels—**into containers** at the end of your laboratory bench. When sinks are used as wastebaskets they may overflow. Do not attempt to unplug a clogged sink; consult your instructor.

10. With respect to what you wear, your shoes should cover your feet to protect them from spilled chemicals or dropped objects. Short sleeves on shirts and blouses are better than long sleeves. In any case, the sleeves should not be loose and floppy. Many easy-care, synthetic fabrics are ultraflammable and should not be worn in the laboratory. Do not wear dangling necklaces or bracelets.

11. If your **hair** is long, fluffed out, and loaded with chemicals, it is quite flammable. At least pin or tie it back in some way while you work around Bunsen burner flames—yours or other peoples'!

12. Radios, unfortunately, are distracting, and you have to be attentive to what you are doing experimentally. That's why **radios are not allowed** in the laboratory.

13. Every time you select a flask, beaker, cylinder, or test tube for some experiment, examine it for **cracks and broken edges**. Sometimes a broken edge can be tolerated, but under no circumstances use a cracked container.

14. Never taste a chemical. Check odors, if you must, only very cautiously.

15. Fitting glass tubing to a rubber stopper: *WARNING: These directions must be followed in order to prevent a serious and very painful cut to your hand.* The hole in the rubber stopper should be slightly smaller than the outside diameter of the glass tube. Before inserting the tube, moisten both the outside of the tube at the end and the inside of the hole in the stopper. Glycerol ("glycerine") is an excellent lubricant for this purpose, but water works well enough. Lubrication of the stopper and the tubing are mandatory. Also mandatory is the use of a towel for holding the glass as you insert it into the stopper. Twist the tubing and the stopper in opposite directions and gently work the glass tubing into the stopper. To avoid snapping the tubing and creating dangerous jagged edges, hold the glass tubing with the towel close to the stopper. If your hands are close together, even if the glass breaks, it cannot move far enough or vigorously enough to cut your hand through the towel.

16. Always turn off a Bunsen burner if you are not using it, and never leave one on that is unattended.

17. Never pour water into concentrated acid when mixing these substances.

18. Remember, there are three ways by which **toxic chemicals** can enter your body; (1) by the respiratory tract, (2) by the mouth, and (3) by the skin, especially if the substance is lipophyllic (fat-soluble).

19. Be careful to follow any additional precautions set forth in the Experiment directions.

Protecting Our Environment:
LABORATORY WASTES

General Principles. Each Experiment will have instructions pertaining to any unusual disposal that falls outside these general guidelines. We wish to emphasize that anything toxic placed into our sewage systems must be removed by the municipality, at taxpayers expense, before returning the water to the natural water cycle. In addition, careless disposal and/or illegal disposal of wastes, is morally repugnant, and can harm us and the next generation as well.

1. Never put solid waste materials in the sinks. Put such materials in solid waste containers or special receptacles designated for that purpose.

2. Always pour aqueous liquids directly into the drain rather than into the bottom of the sink where some of the solution (containing solute) may remain and splash onto desks or clothes or skin when the sink faucet is turned on.

3. Water solutions of some substances, for example "heavy metal ions," cannot be disposed of in drains, according to the laws of some states, and should not be in any case. You must follow the directions of you instructor in these matters. Environmental protection is the responsibility of all of us.

4. If the waste consists of liquid above a solid, decant the liquid into the drain hole in the sink (see Fig. 6) and put the solid into the designated container.

5. Never pour nonaqueous solutions into a sink unless instructed to do so. There will be a container designated for these discards, perhaps in a fume hood.

Important Techniques and Common Laboratory Operations: CARE OF LABORATORY EQUIPMENT

General Rules. The efficiency of your lab work will be markedly improved by having your equipment clean, dry, and in order at the start of the period—in other words, leaving it properly cared for at the end of each period. All metal equipment, including the wire gauze, should be clean and dry before returning it to storage; this prevents rusting.

Washing Glassware. Washing glassware is not too onerous a chore if a few simple rules are followed:

1. Rinse the article and add a few drops of liquid detergent (or a sprinkle of solid detergent if that is provided instead).

2. Select a brush that will fit into the article without cracking the neck opening. Wet the brush and, without adding more water to the glass, scrub with the brush, especially where the article is dirtiest and around the rim.

3. Rinse the article and repeat the detergent treatment if any areas are still stained or greasy; sometimes solid detergent is needed for difficult stains. If that fails, ask the instructor.

4. Finally, rinse the article thoroughly inside and out (including the hand holding the article) and turn the article upside down to drain. Low-form glassware such as beakers and Erlenmeyer flasks can be inverted on pegboards or, if necessary, on a paper towel laid on the desk. Test tubes and graduated cylinders should not be stood on end but should be put on pegs or placed upside down in a test tube rack. If left upright, a pocket of salts (or even detergent) may accumulate in the bottom after drying, which can affect the result of the next experiment! If glass is well washed, rinsed thoroughly, and dried upside down, the salts from city water will not adhere, and the glass will dry "spotless."

Important Techniques and Safe Operation of Laboratory Equipment: THE BURNER

The most common laboratory burners are Bunsen and Tirrill burners (Fig. 2).

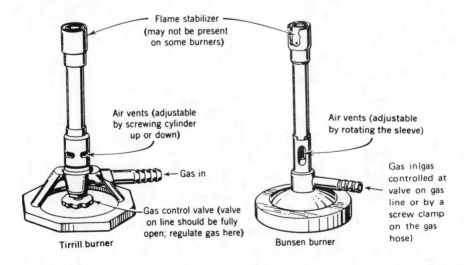

Fig. 2. Laboratory Burners.

The Tirrill burner has a gas-control valve. When the Tirrill burner is used, the main gas valve at the line should be opened fully, and the rate of gas flow should be regulated with the gas-control valve on the burner (Fig. 2). Gas flow in Bunsen burners must be controlled at the main gas valve (or by means of a Hofmann screw clamp on the flexible gas tubing). Five volumes of oxygen are needed to burn I volume of propane to carbon dioxide and water. Since air is only 20% oxygen, 25 volumes of air are needed to burn 1 volume of propane. If the gas has only the air surrounding it when it leaves the top of the burner, much of the gas will be wasted; it will be incompletely burned. Carbon and carbon monoxide, instead of carbon dioxide, will be produced. If air is drawn in through the air vents at the bottom of the burner, it will mix with the gas, and combustion can be complete at the top of the burner (Fig. 2). If these air vents are shut, the flame will be "bushy," sooty, and orange-yellow at the tip (Fig. 3). If the air vents are wide open and if the gas pressure is high, the flame will tend to separate from the top of the burner, and it may even blow itself outTo obtain the hottest flame and the most efficient heating, simultaneously adjust the gas valve and the air inlets until the flame has an inner blue cone and no orange-yellow tip, the flame makes a slight roaring sound, and it stands 5 to 6 in. tall. The intensity of the flame may be reduced by simultaneously closing (not necessarily by the same amount) the gas valve and the air inlets. By trial and error you will learn to produce the burner's hottest flame and then to soften it without creating a bushy, sooty flame. If the air vent is kept wide open but the gas supply is cut back, the rate of propaga-

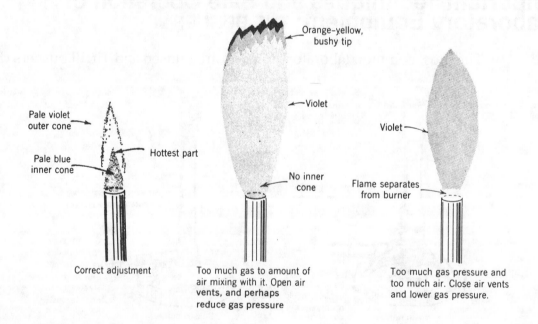

Fig. 3. The adjustment of laboratory burners is done until the best flame appearance is achieved.

tion backward of the flame will be greater than the rate of flow of gas and air, and the flame will move downward through the burner ("strike back"). This may not be noticed and can have serious consequences if the burner becomes hot enough at the base to melt and ignite the rubber tubing.

Heating Liquids and Solids. Organic liquids (e.g., gasoline, ether, alcohol) must never be heated over open flames; they are low-boiling and extremely flammable. The vapors of chlorinated hydrocarbons such as chloroform and carbon tetrachloride form poisonous gases in the presence of free flames or hot metals. To heat water or an aqueous solution in a flat-bottomed flask or beaker,

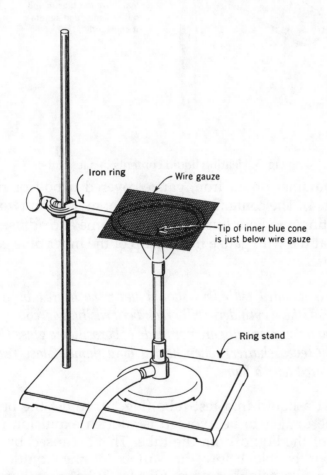

Fig. 4. When you heat a flask or beaker on a wire gauze, you will get most rapid heating when the surface of the gauze cuts across the tip of the inner blue cone of the flame. For a slower rate of heating, either adjust the burner to make that blue cone less intense or raise the iron ring holding the gauze.

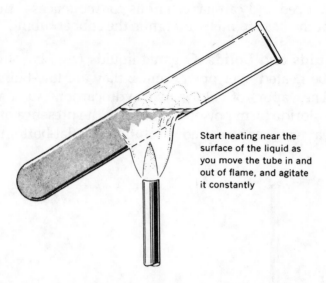

Start heating near the surface of the liquid as you move the tube in and out of flame, and agitate it constantly

Fig. 5. Heating liquid contents in a test tube.

set the container on an iron, gauze-covered tripod or ring affixed to a ring stand (Fig. 4). The container should be not more than two-thirds full. For most rapid heating, use your burner's hottest flame and adjust the height of the iron ring so that the wire gauze hits the tip of the inner blue cone—the hottest part of the flame.

Warning: Graduated cylinders should never be heated by a flame. Because of the etched marks, these cylinders will crack very easily. Moreover, a number of graduated cylinders sold commercially are not made of borosilicate glass (Pyrex, Kimax) and thus break easily (even shatter) when heated by a flame. Glass reagent bottles must also never be heated with a flame.

To heat a liquid in a test tube (Fig. 5), adjust the burner to give a softer flame. The danger to be avoided is sudden expulsion ("bumping") of large amounts of the liquid from the tube. This is caused by rapid formation of a large vapor bubble below the surface of the liquid. To prevent scalding yourself or your neighbor, never fill a test tube more than two-thirds full. Hold the tube in a test tube holder and shake the tube gently as it is being heated.

Warning: Never heat the bottom of the tube, and never point it toward yourself or a neighbor.

Start heating the liquid at a short distance below its surface (Fig. 5), constantly agitating the tube and moving it in and out of the flame. It helps to insert into the tube a long stirring rod which you rotate briskly while the tube is heated. When the liquid is about to boil, remove the tube from the flame and only with great caution return it to the flame. Solids may be heated in Pyrex or Kimax test tubes directly in the flame, noting precautions stated in the experiment. They may also be heated in evaporating dishes or crucibles supported on wire gauze or on clay triangles on rings above the flame if the lab manual or instructor so directs.

Important Techniques and Common Laboratory Operations:
REPORTING DATA

General Suggestions for Prelab Preparation. The experiments in this manual have been carefully selected and designed. Each experiment has a purpose, but it will have value to you only when the objectives are understood. Give these objectives your best time and thought.

Before coming to the laboratory, read the assigned experimental work with these points in mind:

1. What are the objectives? The experiment may have more than one objective. Is it to learn a technique? To demonstrate a chemical law or principle? To make a determination that will characterize a substance?

2. What techniques are used?

3. What equipment is needed?

4. What materials are used? How much? In what concentration? When you go to the reagent table, have in mind whether the sodium hydroxide solution, for example, is 6 M, 2.5 M, or 0.02 M. The success of your experiment may depend on these factors. (The above numbers are units of concentration you will study.) When obtaining the material think about the nature of the substance—acid? base? salt? soluble? insoluble? a reactive ion? flammable? poisonous?

Students may consult each other ("two heads are better than one") and compare experimental results, provided that each is contributing to the information gathered and used and not "saving time"—and brain-power—by copying. Each student must ultimately understand the purpose of an experiment, how that objective was attained and demonstrated, and the evidence that was obtained. There is no collaboration or consultation on a laboratory examination!

The student's notebook is the best source of review of the laboratory work before a lab exam, and in some practical laboratory exams, the use of the student's laboratory reports may be allowed. If the experiment was not originally written up with understanding, or if errors in the original report were not corrected, a lab examination will be difficult.

Organization and Format of the Written Report. In most cases a form is provided for the report. Just answer the questions, put the measurements in the right blanks, make the calculations, and that's it! In other cases, the instructions may be to "Make your own report sheet." In this case, it is well, when reading the experiments over ahead of time, to formulate the report sheet. What are you trying to find? What information do you need to know to get your answer? Arrange these items logically on the page so that the pieces of information, and calculations involving them, lead you step by step to the answer. The most important rule: LABEL ALL NUMBERS with the right unit—grams, milliliters, and so forth.

Numbers that are going to be added or subtracted should be entered with decimals lined up vertically, and for subtracting, the larger number should be entered above the smaller, for example, in recording the masses or volumes in a density determination or volumes in a titration. Note the format for reporting Experiment 2 on density, or Experiment 8 for determining the mol/mol and mass/mass relations in a reaction.

Precision Versus Accuracy. Accurate results are acceptably close to a true, or known value; precise results may agree with each other within an acceptable range but may not necessarily agree with the true or known value. Some examples may help to clarify this.

1.   A pH meter is standardized with a buffer marked pH 7.0; actually this buffer is pH 8.0. Three readings on an unknown solution are taken, and these are 10.0, 10.1, and 9.9. These values would be considered precise, but they would not be accurate because the actual pH of the unknown would have been closer to 11.0 if the proper standard had been used. Another pH meter standardized with a true pH 7.0 buffer might give values of 11.0, 11.5, and 10.5 for the unknown solution. The average of these would be accurate in this case, but there is too much variation to consider them precise.

2.   Three titrations are made of an unknown acid with a "standard" base, using these volumes of base: 26.0 mL, 25.0 mL, and 24.5 mL. These values represent too much variation to be precise. The average of these is 25.2, which would also not be accurate if the correct value was 24.8, for example.

Significant Figures. Over the years, experimentalists have reached agreements on certain ways to report numerical data, ways that signify in the data themselves how precise the measurements are or were. We review here some of these methods as they are normally practiced by chemists. A full treatment would take us into statistics, a subject in itself and far beyond both our time and needs. We survey here only those rules that relate to the kinds of numerical data obtained in some of the experiments in this laboratory text.

Warning: The chief error we try to avoid is that of claiming by the data a precision greater than is possible with a particular measuring device, meter stick, graduated cylinder, pipet or buret, or a laboratory weighing balance. A closely related error comes in not rounding off calculated results correctly.

Significant Figures and Direct Results. A "direct result" is simply a datum obtained by a direct measurement without any calculation. A distance, for example, is reported to be g cm. To a scientist, that number signifies not only something about the distance but also something about the precision of the measurement. The distance is reported as 9 cm, not 9.0 cm or 9.00 cm. The report of "9 cm" signifies that the measuring instrument was so crude that all one can say is that the distance is closer to 9 cm than it is to 8 cm or to 10 cm. With a better instrument, the distance was found to be 9.2 cm, and the report "9.2 cm" means that the distance is closer to 9.2 cm than to 9.1 cm or to 9.3 cm. With a still better instrument the distance could be reported as, for example, 9.25 cm, meaning that both the "9" and the ".2" are known with certainty to be true and that the only uncertainty is in the ".05." The distance is closer to 9.25 cm than to 9.24 cm or to 9.26 cm.

This example illustrates that two things must be considered when we record and report data: the magnitude ("how much?") and the number of digits ("how precise?"). To handle the latter, scientists work with the concept of significant figures; that is, figures that signify something about precision by the count of the figures allowed to remain in the number.

The significant figures in a number consist of all the digits known with certainty to be correct plus the first digit whose value is uncertain. REVIEW AND FOLLOW THE RULES ON THE MEANING AND USE OF SIGNIFICANT FIGURES THAT IS FOUND IN YOUR TEXTBOOK.

To summarize all this advice, let us use as an example Experiment 10, *Hydrates of Copper Sulfate and Calcium Sulfate.* The last item on the report will be: moles of water bound per mole of original substance (name it or give formula). What information do we need to attain this objective? We would need (1) the mass of the sample of plaster of paris, and (2) its formula and formula mass. (3) From these we calculate the moles of the plaster of paris used. (4) We then need the mass of the hydrate formed, (5) its formula and its formula mass. (6) From the difference between the mass of the hydrate formed and the mass of the original sample of plaster of paris, we can calculate the mass of water bound in the hydrate. (7) From the mass of water bound, we calculate the moles of water bound, and (8) from the moles of water bound AND the moles of original plaster of paris, we can get the moles of water bound per mole of original sample. To sum up: in organizing a report (1) organize the data systematically to give you the information needed for the next step, (2) arrange data logically so you can add, or subtract, figures accurately, (3) label all numbers, (4) use only significant figures in your data, and (5) if possible, make your calculations on the same page. If you use a calculator, then set up each solution on the page (with all numbers labelled) and give the calculator result, rounded to the correct significant figures, in the proper place in the report. Make your report a record that someone else—in your absence—can read and learn the facts!

Important Techniques and Common Laboratory Operations:
METHODS OF SEPARATING AN INSOLUBLE SOLID FROM A LIQUID

Decantation. A dense insoluble solid may be separated from the liquid above

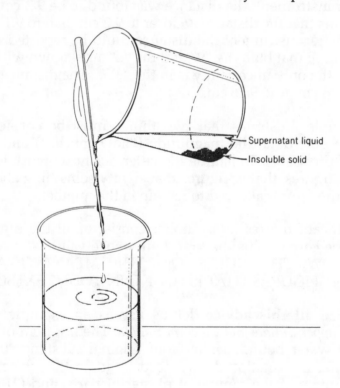

Fig. 6. Decanting a liquid from a solid.

it by simply pouring off—decanting—the liquid (Fig. 6). The solid may be washed repeatedly by adding the wash liquid, stirring the mixture, allowing the solid to settle again, and decanting the wash liquid.

Filtration. The preparation of a filter-paper cone and one way to support a funnel are illustrated in Fig. 7. A small funnel can also be supported in an Erlenmeyer flask or in a test tube in a test tube rack.

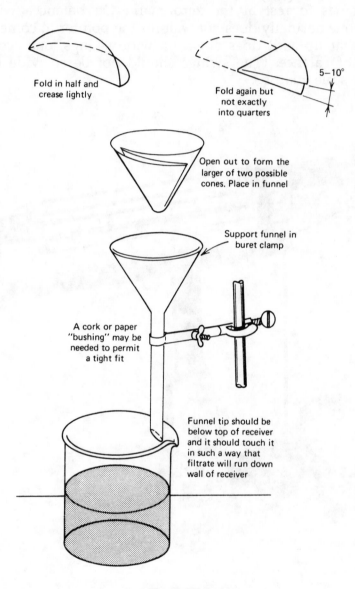

Fold in half and crease lightly

Fold again but not exactly into quarters

5–10°

Open out to form the larger of two possible cones. Place in funnel

Support funnel in buret clamp

A cork or paper "bushing" may be needed to permit a tight fit

Funnel tip should be below top of receiver and it should touch it in such a way that filtrate will run down wall of receiver

Fig. 7. Filtration.

Important Techniques and Common Laboratory Operations: MEASUREMENT OF MASS

Single-Arm Balances. Some of the common types of laboratory balances are shown in Figs. 8 and 9. Each has a beam consisting of two or three notched, weight-equipped scales and one unnotched scale. Each beam has a pointer that swings on either side of a zero mark as the beam swings up and down. On some balances the pointer moves in the field of a magnet which quickly "damps" the swinging. On such balances the scale is balanced when the pointer comes to rest at the zero mark. On balances without magnetic damping, one normally does not wait for the pointer to come to rest. When it swings as far up as it does down on wide swings, one accepts that as the condition of balance. (The swings should not be so wide that the pointer bumps the scale guards.)

Fig. 8. Balance. When a beam is notched, its rider (or balance weight) must hang in a notch—never between notches. On an unnotched beam, readings are taken to the left of the rider, unless the rider has some mark or arrow indicating otherwise.

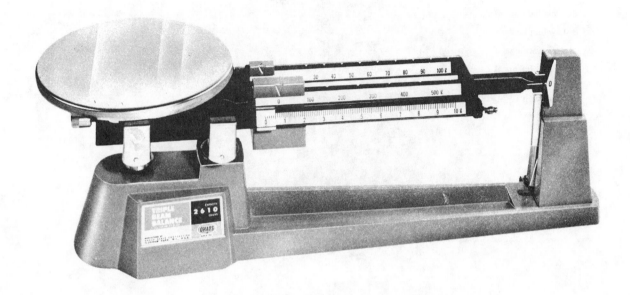

Fig. 9. Platform balance.

When all weights are at their zero positions and the balance pan is empty, the beam should come to zero. If it does not, the balance's special adjustment system should be used to make the needed correction. (This should be supervised by the instructor until you have gained some experience.)

When weighing something, remember that weights on notched scales must always ride in notches.

Top-Loading Balances. Your laboratory may be equipped with one or more top-loading precision balances. Their operating principles are easy to learn, and your instructor will show you exactly how to handle whichever type is furnished in your laboratory. Figures 10 and 11 illustrate one of the many brands of top-loading balances.

① To add weights in whole gram units

② Leveling bubble

③ Scale light on/off

④ Leveling knobs

⑤ Scale zeroing dial

⑥ To add fractional weights

⑦ Tare knob

⑧ Pan

⑨ Scale

⑩ Tare scale

⑪ Case

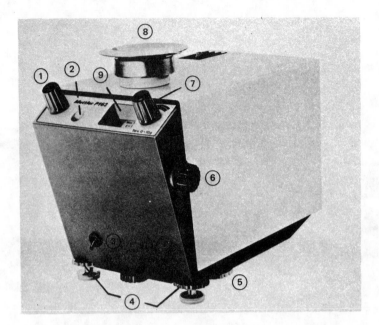

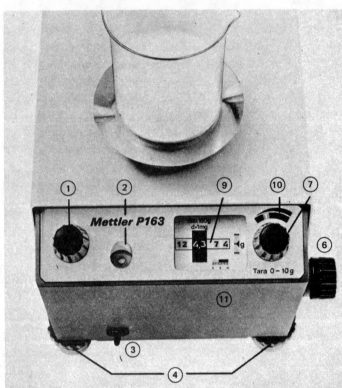

Fig. 10. The Mettler P163 top-loading mechanical balance. The circled numbers are identified on the left. (Photos courtesy Mettler Instrument Corporation.)

16

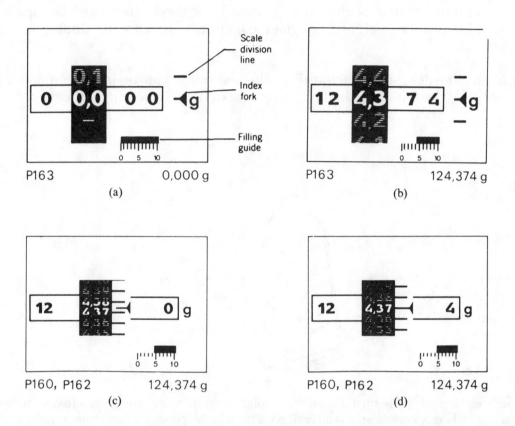

Fig. 11. Views of the readout panels for Mettler top-loading mechanical balances (160-g range). (a) Zero load; tare at zero. Notice the filling guide's left edge is at zero. (b) Here it is exactly balanced with a load of 124.374 g. The filling guide's left edge is between 4 and 5 on its scale. Its edge would move to the right if you added more substance to the weighing pan. Its scale divisions are 1 g each. (c) Older Mettler P160 or P162 balances have a pointer instead of an index fork. Here it is nearly balanced at 124.37 g. (d) It is at exact balance, l24.374 g on P160 or 162 balances. (Photos courtesy Mettler Instrument Corporation.)

USE AND CARE OF THE BALANCE

Weighing Solids. Solid chemicals should never be placed directly on the weighing pan; they could corrode it or spill down into the works of a top-loader and corrode important parts.

Most solids can be weighed on weighing papers or in a weighed dry beaker. Figure 12 shows one way for weighing small amounts of a solid chemical on a triple beam balance (Fig. 8).

Some chemicals must not be weighed on paper. A strong oxidizing agent such as potassium chlorate should be weighed into a glass beaker. Strong

17

alkalies ("lye"), when exposed to humid air, pull moisture from the air so rapidly that the paper under the chemical becomes wet and extremely hazardous because of the caustic property of these chemicals. These alkalies should also be weighed in glass containers. In case of doubt, ask your instructor.

If you do spill a chemical on the weighing pan or in other parts of the balance, clean it immediately. The pans can usually be lifted off easily.

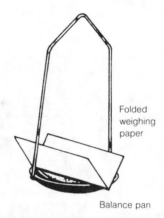

Folded
weighing
paper

Balance pan

Fig. 12. When weighing relatively small amounts of a solid chemical on a balance, use a folded piece of paper like that shown here. (Sometimes it is advantageous to fold the paper a second time at right angles to the first fold to make a slight depression in the center of the paper.) The top edges of the paper can rest against the hanger wires of the balance pan in Fig. 8.

Chemicals should generally be weighed "by difference." Weigh the empty container, then the container with the substance, and then subtract the two weights to get the net mass of the substance. With top-loaders you can often shorten this operation by using the tare knob.

A clean, dry scoopula or other type of spatula is used when transferring small portions of a chemical from its container to the weighing paper (see Fig. 13).

Weighing Liquids. Liquids should never be put directly on the balance pan, but placed instead in a preweighed beaker, Erlenmeyer flask, or graduated cylinder of appropriate size. This container should always be removed from the balance before transferring liquid to, or from, the stock container to avoid dropping the liquid on the balance or balance pan. Accidental spills should be cleaned up immediately.

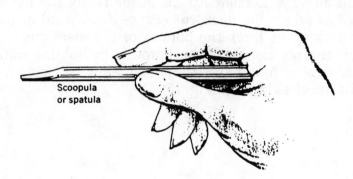

Fig. 13. To manage the addition of small portions of a solid chemical to a container, hold the scoopula or spatula as shown here, and with the index finger tap the scoopula gently.

Important Techniques and Common Laboratory Operations:
MEASUREMENT OF VOLUME

Graduated Cylinders and Pipets. The surface of a liquid or a solution usually curves upward where it meets the walls of the container.

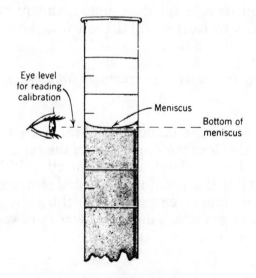

Fig. 14. Reading the bottom of the meniscus.

This crescent-shaped surface is called a meniscus (from a Greek root meaning moon). For best accuracy and reproducibility, graduated cylinders and pipets should be read along a horizontal plane at the bottom of the meniscus (Fig. 14). (In rare cases where the meniscus curves downward at the walls of the container, read along its top.) The bottom of the meniscus can be made to stand out more sharply against the background by holding something darker than the liquid behind it and just below its meniscus. Using a finger will be satisfactory for colorless liquids. (Fig. 15)

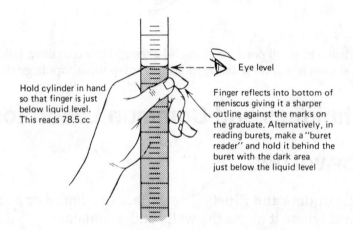

Eye level

Hold cylinder in hand so that finger is just below liquid level. This reads 78.5 cc

Finger reflects into bottom of meniscus giving it a sharper outline against the marks on the graduate. Alternatively, in reading burets, make a "buret reader" and hold it behind the buret with the dark area just below the liquid level

Fig. 15. Reading the meniscus.

Pipet bulbs. A large number of differently designed suction bulbs are available. When you are scheduled for an experiment in which a pipet may be used, your laboratory instructor will demonstrate how to use pipets and how to fill them with a suction bulb.

Warning: Never allow the liquid to be drawn up into the bulb. Serious contamination will result.

Using suction, draw the liquid to be transferred up above the desired volume mark, quickly close the upper end of the pipet with a finger, and then let the liquid level drop to the desired point by quickly removing the finger from and returning it to this position. (This skill requires practice.) A droplet at the lower end of the pipet is removed by touching the pipet tip to the interior wall of the container from which the liquid was removed (or to any clean glass surface).

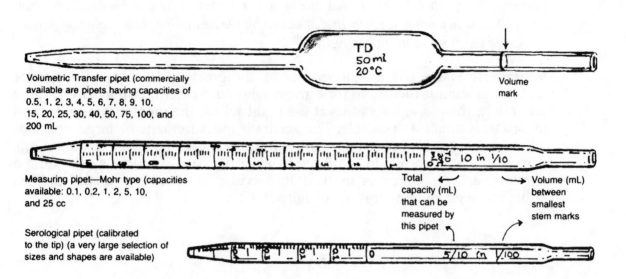

Volumetric Transfer pipet (commercially
available are pipets having capacities of
0.5, 1, 2, 3, 4, 5, 6, 7, 8, 9, 10,
15, 20, 25, 30, 40, 50, 75, 100, and
200 mL

Measuring pipet—Mohr type (capacities
available: 0.1, 0.2, 1, 2, 5, 10,
and 25 cc

Serological pipet (calibrated
to the tip) (a very large selection of
sizes and shapes are available)

TD
50 ml
20°C

Volume
mark

Total
capacity (mL)
that can be
measured by
this pipet

Volume (mL)
between
smallest
stem marks

Fig. 16. Volumetric pipets.

Calibrated Beakers and Flasks. Rough measurements of volumes are possible
with commercially available graduated beakers and flasks. Remember, their
accuracy is limited because as you increase the radius, you greatly increase the
volume for each increment of the height.

Calibrated Test Tubes and Medicine Droppers. An ordinary 6-in. test tube
(20x150 mm) can serve as a crude "graduate" for rough measurements if it is
first calibrated. Trace the outline of a test tube on the outside back cover of the
laboratory manual. Using a pipet or a graduated cylinder, transfer 1mL of
water into the tube and mark its level on your drawing. Similarly, mark 2-mL,
3-mL, 5-mL, 10-mL, and 15-mL levels. When an experiment does not specify
precise volumes, this outline of the test tube will enable you to mark a
particular volume by the location of your finger or thumb as you grasp the test
tube.

A medicine dropper may be used to transfer fairly accurately known
volumes of water or of dilute solutions, if the number of drops per milliliter of
water is first determined. (This number varies widely with different liquids.)
Calibrate one of your medicine droppers in this way and reserve it for such
volumetric transfers. You can make a small scratch with a triangular file for
your calibration marks. Another way to calibrate a dropper pipet is to mark
the height of the liquid when you have removed the desired volume of liquid
from a 10-mL graduated cylinder. Some dropper bulbs are made in such a way
that completely squeezing all air from them will allow a specified volume of
liquid to be drawn in (see Experiment 1).

Warning: Remember, however, that one of your own (even calibrated) droppers must never be used in a reagent bottle that is shared by the rest of the class. Serious contamination of the stock solution can result.

Volumetric Flasks. As mentioned above, the greater the radius, the greater will be the volume measured for a given height increment. To measure volume accurately, therefore, the radius at the point where the meniscus is to be read must be as small as possible. For accurate measurement of large volumes when making solutions whose concentrations are to be exact, a volumetric flask is used. As shown in Fig. 17, this has a flat base, a bulbous body, and a narrow neck on which is marked the level corresponding to the desired volume at a given temperature, usually 20°C.

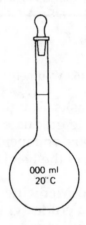

000 ml
20" C

Fig. 17. Volumetric flask.

Important Techniques and Common Laboratory Operations:
MAKING SOLUTIONS, INCLUDING DILUTIONS

As you will learn by consulting your text, there are many types of liquid solutions. Dilute or concentrated, percent concentrations by mass, percent concentration by volume, molar and normal are among the most common types you will be using or making. Liquid solutions consist of a liquid continuous phase, the solvent, and a solute, which is a solid, liquid, or gas dispersed in the solvent. The amount of solute in relation to the final volume or final mass of solution determines the concentration.

Preparation of Solutions. Once the desired amounts of solute and solution have been calculated for the particular concentration wanted, certain procedures must be followed to prepare solutions.

1. Stir the solvent while adding the solute.

2. Start with a container large enough to permit stirring and to make subsequent pouring easy.

3. When using concentrated acids, always *add acid to the water* slowly while stirring. With some acids, particularly concentrated sulfuric acid, much heat evolves, and adding water to the sulfuric acid would produce too much heat for absorption by the liquid. The water would boil violently, spraying dilute sulfuric acid on the surroundings, which includes you.

4. Add a solid in the form of powder instead of chunks to a liquid to speed up solution.

5. Sometimes heat (hot water) can be used to speed solution, but heat should be avoided if it might tend to change the properties of the solute. Albumin (egg white), for example, is rendered insoluble (poached) by hot water. "Soluble" starch, on the other hand, must be stirred into a viscous slurry in cold water to separate the granules, then poured carefully into boiling water. Otherwise you will create a "lumpy gravy" mixture.

6. If the solute is gelatinous, or consists of sticky lumps, work a little solvent at a time into the semisolid mass until it becomes a uniform fluid. Then add the remainder of the solvent. If this type of material were added to the bulk of the solvent all at once, the lumps of solute would be hard to chase, trap, and crush with the stirring rod in order to disperse them into the solvent.

Making Dilutions. Dilute solutions of known concentration can be easily made from concentrated "stock" solutions kept in reserve for this purpose.

Suppose you need 250 mL of isotonic (0.9%) sodium chloride, and a stock solution of 5.0% sodium chloride is available. What is the simplest procedure? A 0.9% solution contains 0.90 g of solute per 100 mL of solution, or 0.0090 g/mL. Therefore, 250 mL of a 0.9% solution contains 250 x 0.0090, or 2.25 g of sodium chloride. What volume of 5.0% solution contains 2.25 g of solute? Since each milliliter of 5.0% solution contains 0.050 g of solute, we calculate 2.25 g x 1 mL/0.050 g = 45.0 mL. We would dilute 45.0 mL of 5.0% solution to 250 mL total volume. Diluting (adding water, in this case) a solution does not change the total amount of solute. It merely means that fewer particles of the solute are in one unit of volume of the diluted solution.

A simple method of calculating these dilutions is to use this formula developed for titrations:

initial concentration x initial volume = final concentration x final volume

or

$$C_i V_i = C_f V$$

For the above problem, we would have

$$5.0\% \times V_i = 0.90\% \times 250mL$$

$$0.18 \times 250 \text{ mL} = 45 \text{ mL}$$

Initial and final concentrations must be in the same units, for example, mol/L, g/100 mL, or (%).

Again, we would dilute 45.0 mL of our stock 5.0% solution to 250 mL total volume in an appropriate container. For greatest accuracy, a volumetric flask would be used.

This relation between initial and final concentrations and volumes can be used for dilutions of molar solutions and percent-volume solutions provided that the initial and final concentrations have the same label (e.g., %, M, etc.). For conversion of percent concentration to molar or normal, or vice versa, your text should be consulted.

SUGGESTED APPARATUS FOR STUDENT DESKS

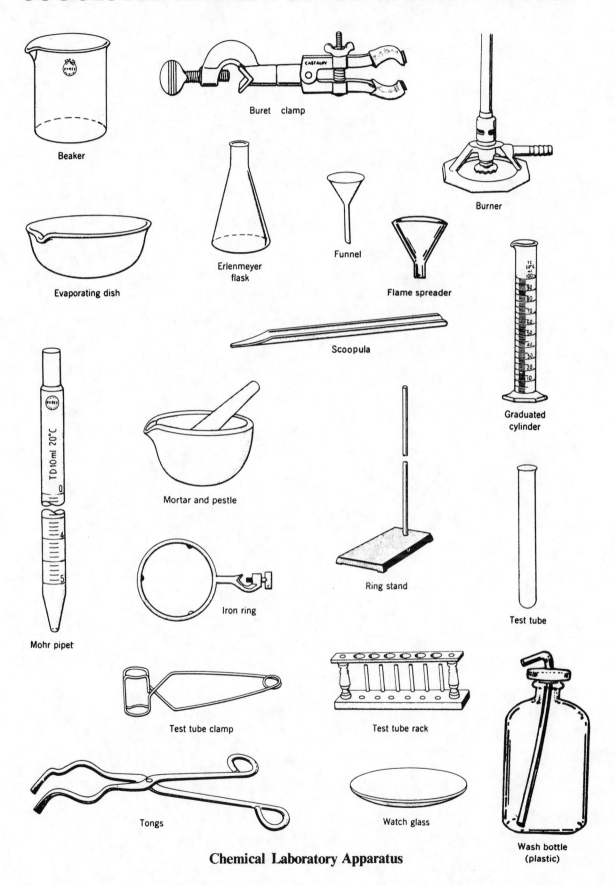

Beaker

Buret clamp

Burner

Evaporating dish

Erlenmeyer flask

Funnel

Flame spreader

Scoopula

Graduated cylinder

Mohr pipet

Mortar and pestle

Iron ring

Ring stand

Test tube

Test tube clamp

Test tube rack

Tongs

Watch glass

Wash bottle (plastic)

Chemical Laboratory Apparatus

Name _____ Name _____

Section _____ Drawer/Lock No. _____ Date _____

SUGGESTED APPARATUS FOR STUDENT DESKS

Check IN	OUT	Equipment for individual use	How many?	Check IN	OUT	Equipment available at each work space for general use	How many?
		Beaker, 50 mL	1			Test tube rack	1
		Beaker, 100 mL	1			Burner, Tirrill or Bunsen	1
		Beaker, 150 mL	1			Ring stand	1
		Beaker, 250 mL	1			Rings (and clamps)	2
		Beaker, 400 mL	1			Buret clamp or utility	
		Beaker, 600 mL	1			clamp	1
		Desk key	1			Wire gauze	1
		Evaporating dish, 10 cm porcelain	1			Crucible tongs	1
						Matches or flint-lighter	
		Flask, erlenmeyer, 50 mL	1			Beaker tongs	1
		Flask, erlenmeyer, 200 or 250 mL	1			Triangular file	1
						Test tube brush	1
		Funnel, about 7.5 cm	1			Sponge	1
		Graduated cylinder, 10 mL	1				
		Graduated cylinder, 50 mL	1				
		Stirring rod, 18-20 cm	2				
		Test tubes (20 x 150 mm)	8				
		Test tube holder	1				
		Watch glass, 10 cm	1				
		Safety glasses					
		Filter paper, 11 cm	1 box				
		Litmus paper, neutral, or blue and red	1 tube each				
		Medicine dropper(s) (number specified by instructor)					

Each person is to read and sign the statement on the reverse side.

Some items, such as thermometers, scoopulas or spatulas, mortars and pestles, and pipets, are available from the side shelf or the stockroom as needed for individual experiments. Return them clean and promptly.

Check-in Procedure. Check the equipment list against the items in your desk. Broken, cracked, or chipped items should be replaced.

Care and Safety Agreement. (Be prepared to sign the following statement and give it to the instructor.)

"I have received the items checked. I accept responsibility for their care, safe-keeping, and proper use. I have read all of the laboratory safety rules and I accept the responsibility for observing all of them. I shall report accidents promptly to the instructor or laboratory assistant."

Signature _____Date_____

Signature _____Date_____

Check-out Procedure

Checking out of the laboratory may count as one experiment. You will be graded on how thoroughly you complete the following items:

1. Clean your portion of the desk top, if necessary scrubbing it with soap. Rinse it. Clean the sink at your desk and the sink at the end.

2. Clean and dry all apparatus. Leave no marks or labels on glassware other than unavoidable scratches and etches.

3. Oil the screws on the clamp and iron ring.

4. Empty the desk completely and line the drawer with a paper towel. Arrange the apparatus to be checked in on top of your desk in approximately the order in which they appear on the check-in sheet.

5. Place all chipped or cracked items to one side. Discard (or otherwise segregate) nonreturnables.

7. Make a list of all missing returnable equipment, after checking with neighbors first to see if they have extras. Then obtain missing items.

8. Return safety glasses if you were issued them.

9. Return proper items to desk drawer as you check in.

10. Lock desk and return key.

11. Additional clean-up assignment:_____

_____

CHAPTER 2   Goals, Methods, and
             Measurements

EXPERIMENT 1 THE MEASUREMENT OF VOLUME

Before starting, read carefully **Volumetric Measurements** and **Reporting Data** in the section on **Important Techniques**.

1 A. METRIC VERSUS ENGLISH UNITS OF VOLUME (OPTIONAL)

If you are unfamiliar with the common metric units of volume and their subdivisions, as well as their relation to our common English units, carry out the measurements listed on the Report Sheet. Only metric volumes will be used in this course. When measuring relatively small volumes such as teaspoonfuls and tablespoonfuls into anything larger than a 10-mL graduated cylinder, it is best to obtain the total volume of at least five of the small units and average them to get the volume of one unit.

1 B. THE CONCEPT OF PRECISION IN VOLUME MEASUREMENT

Illustrate for yourself that graduated beakers and flasks (those with volume measurements etched onto them) are marked only for approximate volumes by measuring some against graduated cylinders.

Equipment. You will need a 50 (or 100) mL graduated cylinder, a 125 mL graduated erlenmeyer flask, and two graduated beakers, 100 (or 150) mL and 250 mL.

Procedure. Into the flask and both beakers, put water up to the 50 mL mark, as carefully as you can determine. Remember to read the bottom of the meniscus as the point to line up with the mark. Then for each, transfer all of the water to a dry large graduated cylinder. For each, record the volume as read in the graduated cylinder. This exercise will continue after you become proficient with pipets.

1 C. USE OF A PIPET

Obtain a pipet and practice using it according to the instructions. When filling a pipet (Fig. 18), the tapered end is held beneath the surface of the liquid at all times. The liquid is then drawn into the pipet by suction until its level is at least equal (or greater than) the volume of liquid to be delivered. (Note that the values on the Mohr pipet increase downward. In no case must the liquid in the lower end of the pipet below the final volume mark be used, since it cannot be measured.) In a transfer pipet, however, the liquid should be drawn to the mark in the neck and all the liquid delivered. Adjust the level of the liquid in the pipet with your forefinger or with the suction device.

Use a pipet bulb when filling a pipet, and follow the instructions of your laboratory instructor in all cases.

WARNING: Never draw liquid into the suction device. Never use your mouth as a suction device.

For greatest precision, use the smallest pipet available for the volume of liquid desired.

Procedure. After obtaining a pipet (preferably have both a 10 mL Mohr pipet and a 10 mL volumetric pipet), and a suction bulb, practice drawing water into the pipet and delivering it, until you can do this easily. With the Mohr pipet, practice delivering 1, 3, 5, and 8 mL of water (for example,) into the graduated cylinder.

When you are proficient in using the pipets, carefully fill a 10 mL pipet and deliver the 10 mL of water to a 10 mL graduated cylinder. Read and record the volume. If possible, do this with the transfer pipet.

1 D. EFFECT OF DIAMETER OF CONTAINER ON PRECISION IN MEASURING VOLUME

Procedure. Fill your 10-mL graduated cylinder to the 10.0-mL mark, and a 100-mL graduated beaker to that mark. Using a 1-mL pipet, or your calibrated dropper pipet, remove 1 mL from each of these containers. Note the new volume of the graduated cylinder. How closely does the new volume reflect the removal of precisely 1 mL?

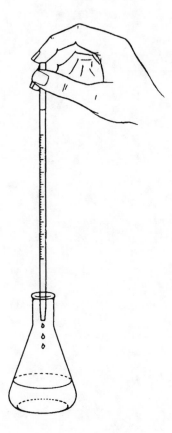

Fig. 18. Technique for pipeting.

EXPERIMENT 1 THE MEASUREMENT OF VOLUME

1 A. METRIC VERSUS ENGLISH UNITS OF VOLUME

Container, filled to the mark = ? metric volume

1-quart measure = _____ mL

1-cup measure = _____ mL

1-ounce measure = _____ mL

1 tablespoon measure = _____ mL

1 teaspoon measure = _____ mL

From your observations, calculate how many ounces and also how many teaspoons = 1 liter. (Set up the problems and do the calculations on the back of the page. Use decimals rather than fractions to express numbers less than 1. Watch significant figures.)

_____ Calc. ounce = 1 liter

_____ Calc. teaspoon = 1 liter

1 B. THE CONCEPT OF PRECISION IN VOLUME MEASUREMENT

Container, Filled to 50 mL mark = Volume, in graduated cylinder:

100 (or 150) mL beaker = _____ mL

250 mL beaker = _____mL

125 mL erlenmeyer flask = _____mL

1 C. USE OF A PIPET

10 mL Mohr pipet = _____ mL in the graduated cylinder.

10 mL transfer pipet = _____ mL in the graduated cylinder.

1 D. EFFECT OF DIAMETER ON PRECISION IN MEASURING VOLUME

In which container does the liquid show the greatest change in depth—the 10-mL cylinder with small diameter or the 100-mL beaker with much greater diameter?

EXPERIMENT 2 THE MEASUREMENT OF MASS (WEIGHT); DETERMINATION OF DENSITY

One of the purposes of this experiment is to learn to use various laboratory balances. Your instructor will demonstrate the use of each balance and discuss the precision of each. Before starting this experiment, read **Measurement of Mass** in the section on **Important Techniques.**

Your practice will consist of determining densities of liquids and water-insoluble solids. The experiment will also give you practice in significant figures. Density is a physical property of matter, and it has the dimensions of mass per unit of volume. In the metric system, density is usually given in units of grams per milliliter (except for gases where the units are grams per liter). In the system used in the United States the units of density are pounds per cubic foot or pounds per gallon. With units of mass per volume we must measure both the volume and the mass of that exact volume.

2 A. DENSITY OF A WATER-INSOLUBLE SOLID

For solids that sink in water and do not dissolve, we have a convenient way of measuring volumes as shown in Fig. 19. The apparent change in volume of the water in the graduated cylinder when the pieces of solid are added is their actual volume. One problem is that a graduated cylinder does not permit great precision in measuring volume. We minimize that problem by working with volume changes as large as practical. For example, if the actual uncertainty in measuring volume in a 50-mL graduated cylinder is 0.3 mL and we use a volume of only 1.0 mL the relative uncertainty is 0.30—very high. However, if we increase the volume to 5.0 mL we have decreased the relative uncertainty to 0.06 (0.3 ÷ 5.0), which is better, but increasing the volume to at least 15 mL gives 0.02 (0.3 ÷ 15) for the relative uncertainty. If we then find a density of 5.5555 units (using a calculator), we will have to round to the tenths position, giving us 5.6 g/mL—only 2 significant figures. (The position of uncertainty is in the tenths position since 5.555 x 0.02 = 0.111.) From this it is apparent that we need to use the greatest practical volume of solid. (If we had used only 1.0 mL, we would have had to round to the units position, to 6 g/mL, which is too inaccurate *and* imprecise.) We could use a larger graduated cylinder and get a larger volume, but this is not practical for use of "unknown" materials by a large class.

Be sure to record all values as you get them—don't try to "remember" and record them later!

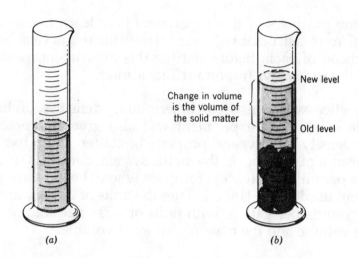

New level

Change in volume
is the volume of
the solid matter

Old level

(a)   (b)

Fig. 19. Determining the volume of an irregularly shaped solid. (a) Liquid level before adding the solid matter. (b) Liquid level after adding the solid matter.

Procedure. To your 50 mL graduated cylinder add 25 to 30 mL of tap water, read the volume as precisely as possible, and record it. Weigh the graduated cylinder plus the water to 0.1 g, and record this mass. Now CAREFULLY, so as not to splash water out of the graduated cylinder, add pieces of the "known" solid. Tilt the graduated cylinder and slide the pieces down the inside. Add pieces of the solid until you have increased the total volume of the solid plus liquid by 15 to 20 mL, but be sure not to add so much solid that the total volume exceeds the 50 mL mark. Also be sure that all of the solid is beneath the water level.

Record the new volume, then weigh the whole assembly and record that figure. Now, by subtraction, you have the mass and volume of the solid and can calculate the density. See how well it agrees with the known value. Calculate your percent error. Your error will be the difference between the accepted value and the value you obtained—it may be positive or negative. If you divide your error by the correct value and multiply that result by 100, you will have your percent error. If this is greater than 5%, check all your calculations, and if necessary repeat the determination.

Now determine the density of one of the "unknown" solids available or assigned. Record your measurements as you make them. Do your calculations. Set up the problems carefully and label all units.

2 B. DENSITY OF A LIQUID

In this experiment, you will determine the mass per unit of volume of a liquid by using a 10-mL graduated cylinder whose uncertainty in reading will be 0.1 mL. As before, it will be to your advantage to use the largest practical volume of liquid so as to have the lowest relative uncertainty for the volume measured. Since you will use a balance that can be read to 0.01 g, the measured volume will have a higher relative uncertainty than the mass. Every number must be labeled with the correct units, and should have the proper number of significant figures.

Procedure. Weigh to 0.01g, a clean, dry, 10-mL graduated cylinder and record the value. Using a pipet, or dropping pipet, add to the cylinder between 7.5 and 10.0 mL of distilled water at room temperature. Read and record the volume to 0.1 mL, weigh the cylinder plus water, and record the value. From your data, calculate the density of water.

Dry your 10-mL graduated cylinder according to the directions of the instructor, and repeat the experiment with one of the "unknown" liquids offered or assigned. Again calculate the percent error for your two determinations. (Do these determinations as precisely as possible not only to be prepared for a practical laboratory examination but also for better performance in other experimental work.)

Name _____ Partner _____

Section _____ Date _____ Due Date _____ Score _____

EXPERIMENT 2 THE MEASUREMENT OF MASS; DETERMINATION OF DENSITY

2 A. DENSITY OF A WATER-INSOLUBLE SOLID

<u>Known:</u>   <u>Unknown:</u>

_____ ; _____ Mass of cylinder + water + solid

_____ ; _____ Mass of cylinder + water

_____ ; _____ Mass of solid

_____ ; _____ Volume of water + solid in cylinder

_____ ; _____ Volume of water originally in cylinder

_____ ; _____ Volume of solid used

Show calculations on back of page.

Calculated density of solid = _____ ; _____

Accepted density of solid = _____ ; _____

% Error in determination = _____ ; _____

2 B. DENSITY OF LIQUIDS

<u>Known:</u>   <u>Unknown:</u>

_____ ; _____ Mass of graduated cylinder + liquid

_____ ; _____ Mass of graduated cylinder

_____ ; _____ Mass of liquid used

_____ ; _____ Volume of liquid used

Show calculations on back of page.

Calculated density of liquid = _____ ; _____

Accepted density of liquid = _____ ; _____

% Error in determination = _____ ; _____

Problems

1. Suppose that when you added the solid to the water in the graduated cylinder, 0.5 mL of air was trapped among the pieces. Which aspect, mass or volume, would this have affected? How would this have affected your calculated density? (In order to answer this question, recalculate what your density would have been.)

2. Suppose you used a wet 10-mL graduated cylinder containing 0.05 mL of water. Which aspect, the mass or the measured volume, would this have affected? Would this have made the calculated density of your unknown liquid higher or lower?

Give the new calculated density:

EXPERIMENT 3 HEAT ENERGY AND CHEMICAL CHANGES

In this experiment you will investigate the role of heat in a chemical reaction, either as a reactant (something added) or as a product (something given off). Although heat is not a "substance," it would appear on one side of the arrow or the other in the chemical equations representing reactions 1-5 below.

Equipment. You will need several large test tubes, a test tube holder, a thermometer, an evaporating dish, crucible tongs or a large metal forceps, and red litmus paper.

Procedure. Carry out the following operations, record your observations, and conclude whether a physical change, a chemical change, or no change occurred.

1. To 5 mL of water (temperature?) in a test tube, add about 0.5 g of ammonium nitrate. Note the temperature immediately.

2. Heat very gently a very small amount of sugar in a test tube, only until you see a change. (Longer heating makes the test tube hard to clean.)

3. Place about 5 mL of 3 M hydrochloric acid in a test tube. Insert a thermometer into the solution and note the temperature. Then add a small piece of metallic zinc. Watch the thermometer and record the highest temperature reached.

4. Obtain 5 mL of 3 M HCl and place it in a test tube. In a second test tube place 5 mL of 3 M NaOH. Record the temperature of each. Pour the acid solution into the NaOH solution and quickly note the temperature again. Record this on the report sheet.

5. Put about 3 mL of distilled water into your evaporating dish and place it on the desk near your burner. Test the water with a strip of litmus paper and note the color. Hold a strip of metallic magnesium securely by means of a crucible tongs (or large metal forceps or metal test tube holder, but not your hand!) and heat the magnesium in a burner flame until ignition occurs. (Do not stare at the intense glare; it may harm your eyes.) Let the ash fall into the water in the evaporating dish and stir it well with your stirring rod. (Afterward, wash the tongs or other holder thoroughly.)

Test the resulting solution with red litmus. If the litmus turns blue, what type of compound is now present in the water?

This new substance, soluble in water and giving a basic reaction, differs from the original shiny magnesium metal.

Name ______________________________ Partner ______________________________

Section __________ Date __________ Due Date ____________ Score ______________

EXPERIMENT 3 HEAT ENERGY AND CHEMICAL CHANGES

Observations: indicate whether a change took place. What did you see?

1. Water and ammonium nitrate:

2. Heat and sugar:

3. Zinc and acid:

Temperature change ______°C - ______°C =______ °C

Calories = 1 cal/°C g x °C x 5 g solution

How many calories of heat were released? = __________ cal

4. Hydrochloric acid and sodium hydroxide:

Initial temperature = _______ ; Final temperature = _________

Change in temperature = _________ ° C

5. Magnesium and heat:

Result of testing solution with litmus:

Questions:

1. In which of the above operations was heat a "reactant?"

2. In which of the operations was heat a "product?"

3. Describe the role of energy in Part 5, in terms of transformation from one type to another during the reaction.

Was more energy given off, or taken in, during the procedure? If more was given off, speculate where it came from.

CHAPTER 3

The Nature of Matter: Compounds and Bonds

REVIEW OF CONCEPTS

Physical properties are characteristics of a substance that can be detected by our senses. These include color, odor, taste, touch, crystallinity, density, and the physical state at a given temperature. If any of these properties change with temperature, but the change is reversed when the temperature returns to its original value, then the change is only physical. The physical appearance of water, for example, changes dramatically if you heat it, starting from its solid form at 0 °C, to its liquid form, then its gaseous form at 100 °C. But simply by cooling the water back to 0 °C, the former states are easily and fully recovered. The water itself has remained water, H_2O.

With a chemical change, new physical properties are observed. To define chemical change, one must consider a substance in terms of its atoms, ions, and molecules, and with respect to electrons, electronic configurations, and bonds. If an observable change can be shown to have involved a rearrangement or redistribution of electrons relative to atomic nuclei, then the change is classed as chemical instead of physical. Although this definition may be mentally satisfying, it is experimentally awkward. Electron rearrangements cannot be seen directly. Enough studies have been done, however, to permit chemists to make generalizations and so enable scientists and others to decide if a change is chemical or physical. The rules do not always work, and there are events at the borderline between the chemical and the physical. However, an event is usually chemical in nature if

1. energy in one or more forms is either given off or absorbed, and if

2. there are observable changes in the physical properties which do not disappear, or reverse, simply by reversing the energy flow or the conditions of the experiment.

EXPERIMENT 4 COMPARISON OF THE PHYSICAL PROPERTIES OF SOME COMPOUNDS WITH THOSE OF THEIR COMPONENT ELEMENTS

On display for you are samples of several elements that will combine with the element chlorine, and form new substances with new physical properties. On the Report Sheet for Experiment 4, record at least two physical properties of chlorine, and two physical properties of each of the other elements on display. For each compound on display, record two physical properties that are different from the properties of the elements from which it was made. The questions on the Report Sheet should help you see the purposes of this exercise.

Warning: Do Not Smell Or Taste Any Substance On Display!

EXPERIMENT 5 EVIDENCE OF CHEMICAL REACTION

Equipment. You will need several clean test tubes, a 10 mL graduated cylinder, and a thermometer. Wash and rinse these well between tests. Dispose of the zinc according to your instructor's directions.

Procedure. Below is a list of substances to be placed in separate test tubes and then mixed. Note the appearance of each substance before, and after mixing. In each case determine the temperature of the solution(s) before and then immediately after mixing. Apply the criteria of chemical change and physical change to each situation; decide whether there was any change at all.

First Test Tube:	Second Test Tube:
1.) 5 mL of 0.1 M $BaCl_2$	5 mL of 0.1 M K_2CrO_4
2.) 5 mL of 0.1 M $BaCl_2$	5 mL of 0.1 M Na_2SO_4
3.) 5 mL of water	1.0 g of NH_4NO_3
4.) 5 mL of 3 M HCl	5 mL of 3 M NaOH
5.) 5 mL of 3 M HCl	1.0 g $NaHCO_3$
6.) 5 mL of vinegar	1.0 g $NaHCO_3$
7.) 5 mL of 3 M HCl	5 mL of 0.1 M K_2CrO_4
8.) 5 mL of 3 M HCl	a piece of zinc metal
9.) 5 mL of 0.1 M $CuSO_4$	a piece of zinc metal
10.) 5 mL of water	1.0 g NaOH (do not touch)

EXPERIMENT 4 COMPARISON OF PHYSICAL PROPERTIES

Element common to all compounds: Chlorine Cl_2

List two physical properties _____

Symbol/ Formula	Name of Element or Compound	Two Physical Properties	
		(1)	(2)

QUESTIONS

1. What physical properties does the "common element" have that are different from those of any of the other elements or compounds on display?

2. If there are elements on display that you could not distinguish solely by their physical properties, list them.

(Questions continue on reverse side.)

3. If there are compounds you could not differentiate from each other by physical properties only, list them.

4. Why were the properties "taste" and "odor" excluded from the properties you investigated?

5. List any compounds that illustrate the law of multiple proportions.

6. For the compounds with color, would you say the color was caused by the ion of the common element or by the metallic element? Explain you answer.

EXPERIMENT 5 EVIDENCE OF REACTION

Combination:	BEFORE: Temperature and appearance	AFTER: Evidence of change
1. $BaCl_2$ + K_2CrO_4		
2. $BaCl_2$ + Na_2SO_4		
3. water + NH_4NO_3		
4. HCl + NaOH		
5. HCl + $NaHCO_3$		
6. vinegar + $NaHCO_3$		
7. HCl + K_2CrO_4		
8. HCl + zinc		
9. $CuSO_4$ + zinc		
10. water + NaOH		

CHAPTER 4 · # Chemical Reactions: Equations and Mass Relations

STOICHIOMETRIC RELATIONSHIPS: MOLES, GRAMS, AND BALANCED EQUATIONS

Stoichiometry is a branch of chemistry concerned with the mass relations between atoms in a molecule and/or compounds in a chemical reaction. It is based on the law of conservation of mass and the law of definite proportions. The law of conservation of mass and energy states that when substances react to form new substances, none of the mass or energy is lost. A balanced equation accounts for all the mass involved in a reaction by showing that all the atoms present in the reactants also appear, rearranged or reorganized, in the products. In the following four experiments we shall check this law, so the measurements must be made carefully. The balanced equation for each reaction is used to calculate how much of each of the reactants should be used. In the first experiment, you will be comparing the ratios of grams of metal consumed and produced with the ratios of moles of the metals consumed and produced. Does the balanced equation represent mole quantities or mass quantities? The second and forth experiments investigate a simple reaction in which the amount of one of the reactants is held constant while the amount of the other reactant is changed. What affect does this have on the amount of product? The third experiment investigates the concept of "water of hydration," illustrating that these molecules are a definite part of the whole molecule's stoichiometry.

EXPERIMENT 6 MASS RELATIONS IN CHEMICAL CHANGES

Chemical changes, or reactions, occur between discrete atoms, ions, or molecules in ratios of small whole numbers, regardless of the masses of the individual particles. All particles of one substance will have the same average mass, but particles of the different substances will have average masses different from each other. Thus, ratios by mass are seldom simple, whole-number ratios. We shall study this fundamental idea in chemistry in this experiment.

Zinc is a more active metal than lead; in other words, it has more of a tendency to lose electrons than lead. If, therefore, a lead ion is brought into contact with neutral zinc atoms and the zinc atoms can leave their electrons with the lead ion. The lead ion will be come a neutral atom, and the zinc atom can now go into solution as a zinc ion. Below is the balanced ionic equation for the reaction, and the coefficients of the reactants are in the simple ratio of 1:1; 1 mole zinc atoms to 1 mole lead ions.

$$Zn + Pb^{2+} \rightarrow Zn^{2+} + Pb$$

To obtain the best results, the masses should be determined carefully to the nearest 0.001 g, which is at least 10 times the precision normally found necessary in this course. If you do not normally weigh to this precision, the instructor will do it for you or make some special arrangement. In any case, when you go to the balance to obtain any of the masses used in this experiment, take your Record Sheet so you can write the mass down on the spot—don't try to remember it to write down later.

Equipment. Obtain a piece of 11 cm filter paper, (and put your name on it in pencil), a large (15 cm) test tube, a zinc strip that is about 5-10 mm by 15 cm long, a small piece of steel wool, 20 mL of a 1.0 M lead acetate solution, and two or three clean, small beakers.

Procedure. Polish the zinc strip with the steel wool very gently, wipe it thoroughly, then find its mass to the nearest 0.001 g. Record this on the Report Sheet.

Put the 20 mL of lead acetate solution into the test tube and stand the zinc strip in the solution, leaving it undisturbed for about an hour. Record the change in appearance of the zinc, which starts immediately.

Recovering the LEAD deposits: After the specified time, gently lift the zinc strip, now heavy with lead deposits from the solution, and transfer it to a beaker of distilled water. Very gently stir the zinc strip in the water, then let it soak for about two minutes. Repeat this process with another supply of distilled water. Be careful not to dislodge the lead! Take the beaker of rinse water, with the zinc strip and lead in it, to the hood. Take along a second clean beaker.

At the hood, gently lift the strip out of the water, and either dip it in beakers of acetone provided, or rinse it with a wash bottle containing acetone. The acetone will displace the water wetting the zinc and lead, and allow them to dry quickly.

Weigh the filter paper (with your name on it!) to the nearest 0.001 g, and record this on your Report Sheet. Use a steel spatula to carefully scrape the lead deposits onto the filter paper. With the still-clean part of the filter paper, wipe as much of the residual deposit off as is possible (see Figure 20). Put the lead and the paper to dry in the place designated by your instructor.

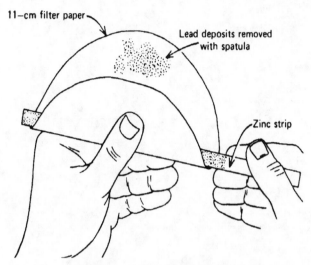

Fig. 20. Removing lead deposits.

Allow the ZINC strip to dry thoroughly, then weigh it to the nearest 0.001 g; record this mass. Now polish it gently with fine steel wool and weigh it again, to the nearest 0.001 g. (Record the mass again.) Any mass lost in this last procedure represents LEAD that you were unable to remove by wiping with the paper, and should be added to the mass of the LEAD on the paper. The drying should be done as quickly as possible (not overnight!), to avoid possible oxidation of the lead.

EXPERIMENT 6 MASS RELATIONS IN CHEMICAL CHANGES

(a) _____ Mass zinc strip, at start

(b) _____ Mass zinc strip after scraping and wiping with paper

(c) _____ Mass zinc strip after buffing

(d) _____ Total mass of zinc dissolved (a - c)

Moles of zinc dissolved = grams of zinc dissolved/f. mass of zinc (e)_____

(f) _____ Mass of lead removed by buffing (b - c)

(g) _____ Mass of paper + lead

(h) _____ Mass of paper

(i) _____ Mass of lead on paper

(j) _____ Total mass of lead formed (f + i)

Moles of lead formed = grams of lead/f. mass of lead (k) _____

1. Using your data: Calculate

(A) The ratio of: grams of lead formed ÷ grams of zinc dissolved = _____

(B) The ratio of: moles of lead formed ÷ moles of zinc dissolved = _____

2. Which comes closer to matching the coefficients of the equation in this experiment, your mass-to-mass ratio, or your mole-to-mole ratio?

3. State briefly why more lead formed, by mass, than zinc disappeared, by mass, from the strip. How many times heavier are lead atoms than zinc atoms?

4. Zinc ions cannot exist in solution by themselves but must be balanced by an equal number of charges on negative ions. What is the negative ion in the solution? Write its name and formula.

5. What is the name of the *salt* formed? The formula?

Error analysis. You have done this experiment only once, and your results are not necessarily as precise as they would be with more practice and with more sophisticated equipment. However, your results should have provided you with an excellent occasion for helping you think about the mole concept and reaching a better understanding of it. *Let us consider how various sources of error would have affected the calculated RATIO of moles of lead to moles of zinc, which should be 1:1.*

Consider the affect of each of these. Because of the error, the measured ratio, "mole Pb to mole Zn," will (check one):

Description of error	Increase	Decrease	No affect
1. Some lead stayed on the zinc strip; it was not fully removed.			
2. The zinc strip was wet when you weighed it before putting it into the lead acetate solution.			
3. The zinc strip was still wet when you weighed it at the end of the experiment.			
4. The lead on the filter paper was wet when you weighed paper + lead.			
5. The zinc strip was in contact with the lead ion solution more than an hour.			
6. Some lead changed to whitish lead oxide while the lead was drying.			
7. Some zinc acetate was trapped in the spongy lead deposit and weighed with the lead.			
8. The lead acetate was not exactly 1 M.			

Which would cause the most error in the moles Pb/moles Zn ratio, zinc *acetate* trapped in the spongy lead deposit, or zinc *oxide* formed by conversion of the acetate during heating? (Hint: The formula for the acetate ion is $CH_3CO_2^-$, so formula mass of zinc acetate= _____? Formula mass. zinc oxide = _____?)

EXPERIMENT 7 THE RELATION BETWEEN MOLES OF REACTANTS AND MOLES OF PRODUCTS (The Reaction of Acid with Magnesium)

This experiment is an excellent classroom demonstration that the amount of product formed is limited by the amount of reactants available. If done as a demonstration, the amounts of reactants specified below should be doubled or tripled. The balanced equation for the reaction tells us that one mole of magnesium and two moles of hydrogen chloride yield one mole of the salt, magnesium chloride, and one mole of the gas hydrogen.

$$Mg + 2HCl \rightarrow MgCl_2 + H_2$$

Determining the mass of magnesium chloride formed would be more complicated than measuring the quantity of hydrogen formed. We may easily trap the gas and measure it, and from the resulting volumes, we can find out if limiting the amount of magnesium or the amount of hydrogen chloride will limit the amount of product formed.

Equipment. You will need three balloons of uniform size, magnesium ribbon, masking tape, 3-125 mL Erlenmeyer flasks, and 150 mL of 1.0 M HCl.

Procedure 1. Weigh a 10.0 cm length of magnesium ribbon. Using its mass, calculate the lengths needed to obtain 0.3, 0.6, and 1.2 g of Mg, in single pieces. Make sure the metal is shiny. The ribbon should be in one piece for each of the three different masses, and should be coiled tightly.

2. Place exactly 50 mL of 1.0 M HCl in each of three clean 125-mL erlenmeyer flasks, using a funnel so that you do not get the necks of the flasks wet with the acid.

3. Carefully drop one of the magnesium metal coils into each balloon. Without allowing the metal to cut the balloon or accidently drop into the flask, fit the rubber balloons over the necks of the flasks and fasten them with masking tape. Be sure no flames are nearby.

4. With your finger on the outside of the balloon, carefully push against the 0.6- g coil of ribbon until it falls into the acid, and observe and record what occurs. Next, push the 0.3-g coil into the acid. Finally, push the 1.2-g coil into the acid. In each case, occasionally swirl the flask, and let it stand until no further reaction occurs. Record your observations and data on the Report Sheet. If the balloon breaks, place the flask in a fume hood (away from flames) until the reaction is complete. Ask your instructor about disposing of wastes.

Name ______________________ Partner ______________________

Section __________ Date __________ Due Date __________ Score __________

EXPERIMENT 7 RELATION BETWEEN MOLES OF REACTANTS AND MOLES OF PRODUCTS

1. Quantitatively estimate the amounts of gas formed in each case, using the amount of gas from the 0.6-g magnesium sample as the standard. Since the three volumes of gas are essentially at the same temperature and pressure, their volumes are directly proportional to their masses (moles), according to Avogadro's principle.

2. How many moles of HCl were used in each flask? ________

3. Calculate the moles of Mg used in each case:

(a) 0.3 g Mg = ________ moles

(b) 0.6 g Mg = ________ moles

(c) 1.2 g Mg = ________ moles

4. Was a significant amount of magnesium left unreacted? If so, in which case?

5. Explain why this should have happened:

6. Calculate the number of moles of hydrogen that should have been generated in each case:

(a) from 0.3 g Mg _____

(b) from 0.6 g Mg _____

(c) from 1.2 g Mg _____

6. What should have been the volume of the hydrogen from (c) at Standard Conditions? (STP)

7. Discuss the comparative volumes of hydrogen gas obtained from the same amount of HCl and the differing moles of Mg.

EXPERIMENT 8 HYDRATES OF COPPER SULFATE AND CALCIUM SULFATE (PLASTER OF PARIS)

The crystals of true *hydrates* appear (and feel) dry, but there are a fixed number of molecules of water for each formula unit of the anhydrous salt. This makes hydrates true compounds because they obey the law of definite proportions. We indicate the attachment of these molecules of "water of hydration" by writing the chemical formula of the substance followed by a raised dot and the water, with its coefficient showing how many are attached. An example is copper sulfate pentahydrate:

$$CuSO_4 \cdot 5H_2O$$

The formula written this way should be understood to be a stoichiometric entity, and its formula mass must include the total mass of water of hydration, as written. In contrast, substances that are *hygroscopic*, or *deliquescent*, attract water from the air in random amounts, eventually forming a solution. You may have observed this in the case of table salt ($NaCl$) and even sugar and could have also noticed it with the calcium chloride ($CaCl_2$) used to melt ice on streets or to dry the air in damp places such as basements.

8 A. Copper Sulfate

Copper sulfate forms a well-known hydrate with a rich blue color,

$$CuSO_4 \cdot 5H_2O$$

The Equipment. You will need a clean, *dry* test tube, bunsen burner, and a medicine dropper. Copper sulfate is a toxic substance, and should be handled carefully.

Procedure. Place a small crystal of this hydrate in a dry test tube and heat it gently. As you do, watch the upper, cooler part of the test tube. On the Report Sheet, note what you observe. Also on the Report Sheet note any change in the appearance of the copper sulfate itself.

Allow the tube to cool. (Tilt it and let tap water run over the outside of the lower part.) Add one drop of water (no more) to the residue and describe what you see.

8 B. Calcium Sulfate

Plaster of paris is also a hydrate because it contains a fixed ratio of salt to water, $(CaSO_4)_2 \cdot H_2O$. Mixed with water it forms a paste or slurry which then hydrates to $CaSO_4 \cdot 2H_2O$. This hardens as excess water evaporates and forms the casts used to support broken bones or to make models of various types. This experiment can be done qualitatively or quantitatively as the instructor designates. These directions are for the quantitative determination of the amount of hydration in the final hydrate. We shall determine the number of moles of water per mole of $CaSO_4$ and see how close it comes to the values represented by the balanced equation:

$$(CaSO_4)_2 \cdot H_2O + 3\,H_2O \rightarrow 2\,CaSO_4 \cdot 2H_2O$$

If you wish to make a mold of a coin or some other object, cover the face of the object with a thin film of petroleum jelly before you press it into the plaster. If you wish to demonstrate why it is necessary to have a cast loose on a limb while it is wet (because it expands on "setting") you can imbed one end of a piece of rubber tubing (about 2 cm) that will just fit around a glass rod in the plaster as you mold it. Then remove the glass rod. When the cast has "set," see how much more tightly the rod fits in the tubing.

Equipment. You will need a 20 x 20 cm square of aluminum foil, waxed paper, or heavy plastic wrap; a 2-3-cm length of rubber tubing, a glass rod that fits the tubing, and a thermometer; if desired, a small object such as a coin; a small spatula, and a small amount of petroleum jelly.

Procedure. Weigh to 0.01g a piece of aluminum foil, waxed paper, or heavy plastic about 20 cm square along with any item you are intending to include in the cast with its coating of petroleum jelly if appropriate. Remove the foil from the balance. Using the 50-mL beaker in the supply of plaster of paris as a scoop, half fill it (about 25 mL) with the plaster and pour the plaster on the aluminum foil. Again weigh the assembly to 0.01g. Be sure to record these masses on the Report Sheet.

Remove the foil and plaster from the balance. Using a small spatula, stir water that is at room temperature into the powder carefully a few drops at a time until you have a moist but not "runny" plastic mass. Do not discard any of the plaster you have weighed. If you used too much water and need more plaster, weigh it carefully in a separate container, and record the mass so it can be added to the mass of the original plaster. Note the temperature of the wet mixture, and also the room temperature.

Place the object you are casting in the plaster, (which may include the rubber tubing), wipe all the plaster from your spatula onto the aluminum foil, and put the assembly inside your drawer until the next lab period (about a week). Weigh it (to 0.01 g) again, calculate the mass of water added to the hardened plaster, the moles of water added, and the moles of water added per mole or original plaster used. Notice the properties of the cast. If possible look at a cross section of a broken cast with a hand lens. Does it seem porous?

EXPERIMENT 8 HYDRATES OF COPPER SULFATE AND CALCIUM SULFATE (PLASTER OF PARIS)

8 A. Hydrates of Copper Sulfate

1. What condenses in the upper of the test tube?

2. Where did it come from?

3. What happened to the appearance of the crystalline copper sulfate?

4. When you added the drop of water to the anhydrous copper sulfate, what did you see? (Did the material appear "wet"?)

5. Write balanced equations for the two reactions you carried out:

(a) Action of heat:

(b) Action of water:

6. Using the information in the equations, explain the changes observed. (Which of the ions, Cu^{2+} or SO_4^{2-}, changed color when it became dehydrated? If possible look at other hydrated salts of SO_4^{2-} and of Cu^{2+}.)

7. Why are hydrates classified as compounds rather than as wet mixtures?

8 B. Hydrates of Calcium Sulfate (Plaster of Paris)

_____ (a) Mass of aluminum foil (or other paper, objects, etc.)

_____ (b) Mass of plaster of paris powder plus foil, etc.

_____ (c) Mass of powdered plaster of paris (b) - (a)

_____ (d) Formula mass of $(CaSO_4)_2 \cdot H_2O$

_____ (e) Moles of $(CaSO_4)_2 \cdot H_2O$

_____ (f) Moles $CaSO_4$

_____ °C Difference between room temperature and the temperature of the wet plaster; heat added or taken in?_____

SECOND LABORATORY PERIOD:

_____ (g) Mass of plaster of paris assembly (now hardened)

_____ (h) Mass of hardened plaster (g) - (a)

_____ (i) Mass of water of hydration (h) - (c)

_____ (j) Formula mass of water

_____ (k) Moles of water of hydration gained

(l) RATIO: Moles H_2O/moles $CaSO_4$. How close does it come to the values represented by the balanced equation?

_____

1. Why is it important to make a plaster cast on a living object loose before it hardens? (What could happen to the leg, arm, etc.?)

2. What were your observations regarding heat given off or taken in as the hydrate formed? What implications does this have for application of a plaster cast?

EXPERIMENT 9 EFFECT OF LIMITING THE CONCENTRATION OF A REACTANT

When sodium carbonate and calcium chloride are combined, a white precipitate forms. How much precipitate will form depends on how much carbonate and how much calcium are available. They react according to the net ionic equation

$$Ca^{2+}_{(aq)} + CO_3^{2-}_{(aq)} \rightarrow CaCO_{3(s)}.$$

Since the ratio of ions needed is one-to-one, we expect equal numbers of moles of calcium to react with equal numbers of moles of carbonate ions. What happens when they are not supplied in a one-to-one ratio? What can we expect to see if one of the ions is more plentiful than the other? Will the amount of precipitate forming in such a case reflect the concentration of the ion in excess, or the ion in short supply? What does it mean to have a "limiting reactant?"

Procedure. Label from 1 to 6 six test tubes of uniform diameter, each with a well-fitted rubber stopper. Place them in a test tube rack.

Using a pipet, carefully add to each tube 5 mL of sodium carbonate of the concentration specified for each tube on the Report Sheet.

Then into each in turn, pipet 5 mL of the calcium chloride solution of the concentration specified for each tube on the Report Sheet.

As you make each addition, close the tube with a rubber stopper and shake it vigorously a uniform number of times (e.g., 20 times). Place the tube in the rack and leave it undisturbed. Note that when you make the addition of 1 M calcium chloride to the 1 M sodium carbonate, the precipitate looks almost gelatinous, but it breaks up into particles when shaken.

When all the tubes have been filled and shaken, record the time and permit the solids to settle undisturbed for about 15 to 30 minutes while you do some of the calculations on the Report Sheet. Make the comparisons at the same time, using tube 3 as a standard.

Name _____ Partner _____

Section _____ Date _____ Due Date _____ Score _____

EXPERIMENT 9 EFFECT OF LIMITING THE CONCENTRATION OF A REACTANT

1. Write a balanced net ionic equation for the reaction between $CaCl_2$ and Na_2CO_3.

2. To get one mole of product, how many moles of $CaCl_2$ are needed? How many millimoles of Na_2CO_3 are needed?

3. How many mmols of Na_2CO_3 and of $CaCl_2$ were in the 5-mL portions of solution specified for each of the six test tubes? Use the back of the Report Sheet for calculations. Record in the Report form below.

4. Assuming 100% reaction, how many millimoles of product could be formed in each of the six solutions? Use the back of the Report Sheet for calculations. Record in the Report form below.

Data and Calculations

Tube No.	Concentration of		Comparative Volume of Precipitate Formed*	Millimoles of Na_2CO_3 in 5 mL of Solution Used	Millimoles of $CaCl_2$ in 5 mL of Solution Used	Calculated Millimoles of Product Possible
	Na_2CO_3	$CaCl_2$				
1						
2						
3						
4						
5						
6						

(*Since the test tubes are the same diameter, the height of the precipitate in each tube gives a direct means of comparing the volumes.)

(Questions continue on next page.)

5. Are your calculated answers proportional to the observed volumes of precipitate?

6. How many millimoles of the following ions will be _left over_ (if any) after the reaction in each tube (assuming 100% reaction)?

Tube	1	2	3	4	5	6
CO_3^{2-}	_____	_____	_____	_____	_____	_____
Ca^{2+}	_____	_____	_____	_____	_____	_____

7. The solubility of $CaCO_3$ at room temperature is about 0.0015 g/100 mL. How many millimoles per liter is this? Show your calculations.

Calculations:

CHAPTER 5 · # Kinetic Theory and Chemical Reactions

EXPERIMENT 10 CHARLES' LAW: THE RELATION BETWEEN VOLUME AND TEMPERATURE AT CONSTANT PRESSURE

Heating a gas increases the average energy and velocity of its molecules, and if there is a chance for them to escape into a larger volume (while keeping the pressure constant) they will do so. Thus a direct relation exists between the volume of a gas and its absolute (Kelvin) temperature at constant pressure. This is Charles' law.

Equipment: You will need
an iron ring and wire screen,
a ring stand,
a clamp (as shown),
a small screw clamp,
a 1 inch piece of glass tubing,
a 250 mL erlenmeyer flask,
a one-holed rubber stopper (to fit the erlenmeyer flask),
a short piece of glass tubing (to fit the rubber stopper),
a large beaker (800 or 1000 mL).

Procedure. Assemble the apparatus shown in Fig. 21, using an 800-mL or 1000-mL beaker and a dry 250-mL Erlenmeyer flask fitted with a one-hole rubber stopper in which there is a short piece of glass tubing that can be closed by a 1-in.-long piece of rubber tubing holding a screw clamp (not a pinch clamp). The flask should be positioned so that as much of it as possible will be submerged when the beaker is filled with water to boiling. Adjust the flame so that the water boils gently, but watch that it does not boil over and extinguish the burner. (The addition of a boiling chip is helpful.)

Continue the boiling for five minutes so that the temperature of the air inside the flask becomes the same as the temperature of the boiling water. Take the temperature of the water and record it as T_i. Put a screw clamp (not a pinch clamp) on the rubber tubing connected to the stopper and fasten it very tightly.

Holding the flask by the clamp by which it is fastened to the ring stand, remove it with the clamp from the boiling water bath. Holding the flask in an inverted position, plunge it into a large reservoir of cold water, conveniently produced by filling a stoppered sink. Stir it around, and after about five minutes, take the temperature of the water in the reservoir. This should be the temperature of the air inside the flask. Record this as T_f on the Report Sheet.

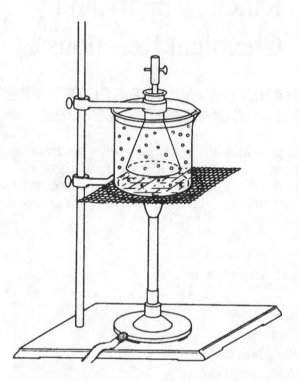

Fig. 21. Apparatus for determining Charles' law.

Holding the flask in an inverted position and keeping the rubber tube under the water, unscrew the screw clamp, allowing the water to replace the air that had escaped as a result of expansion at the higher temperature. Carefully hold the end of the rubber tube tightly closed with your fingers and place the flask upright on the laboratory desk. Loosen the stopper and allow the water to drain from the tube into the flask. Measure the volume of water that was drawn into the flask under the cold water; this equals the volume of air driven from the flask at temperature T_i, (boiling water bath).

To find the amount of air that was in the flask at T_i, fill the flask with water, insert the stopper, and pinch the end of the tubing as you did when removing the flask from the cooling bath. Loosen the stopper, allow the water in the tubing to drain into the flask, and measure the volume of water; record this as V_i.

Complete the calculations and answer the questions on the Report Sheet.

Name ______________________________ Partner ______________________________

Section __________ Date __________ Due Date ____________ Score ______________

EXPERIMENT 10 CHARLES' LAW

T_i = ________ °C = _________ K. V_i = ________ mL

Volume of water drawn into the flask at T_f = ________ mL

T_f = ________ °C = _________ K.

V_f = V_i minus the volume of water drawn into the flask at T_f =

__________ mL

1. Using the formula for Charles' law, and the measured initial volume (Vi), calculate the theoretical final volume,

V_f = _________ mL.

2. What is the percent error of the value you found compared to the calculated theoretical volume?

__________ %

3. Suggest some sources of error you might have made in this determination:

4. Make your calculation for question 1 using Celsius instead of Kelvin temperatures. Are the results significantly different?

EXPERIMENT 11 THE PARTIAL PRESSURE OF OXYGEN IN THE AIR (PERCENT OXYGEN IN AIR)

The behavior of gases of most interest in the health field is the law of partial pressures. In respiration, variations in the partial pressures of oxygen and carbon dioxide determine the exchange of gases at the lungs and at the cells. Each gas present in a mixture of gases contributes to the total pressure exerted by the mixture in proportion to its concentration in the mixture. Thus, in air, the major gases are nitrogen, oxygen, carbon dioxide, and water vapor, and the total atmospheric pressure P_{total} is therefore equal to $P\,N_2 + P\,O_2 + P\,CO_2 + P\,H_2O$. We can measure the total pressure with a barometer. We can learn the partial pressure of water in air saturated with water vapor at any temperature from known data printed in tables or on graphs. By experiments we can determine the percent of oxygen or carbon dioxide in a given gas sample. Nitrogen can then be found by difference. These percents are easily converted into partial pressures. In this experiment we shall measure the partial pressure of oxygen in air.

Equipment. You will need a large (25 x 250 mm) test tube, a small, thin rubber band, a 250 or 400 mL beaker, a thermometer, and iron filings.

Procedure. Slip a very small, thin rubber band over the walls of an extra-large test tube (25 mm x 250 mm), or a 50-mL or 100-mL graduated cylinder. Measure the brimful capacity of this container. Empty the water and sprinkle iron filings onto the wall so as to cover at least half the moistened surface.

Now lower the cylinder slowly and carefully into a 250-mL beaker about one-half full of water. Allow no bubbles to escape from the air now trapped in the cylinder. The amount of air entrapped is equal to the brimful volume of the container. (Expect some water to rise inside the cylinder at the start, however, for the water outside the cylinder exerts a pressure slightly greater than the initial (atmospheric) pressure of the air on the inside.) See Fig. 22.

Record the temperature and the barometric pressure when you start the experiment. Set the apparatus aside, undisturbed, until the next laboratory period. During the intervening time, the oxygen in the entrapped air will react with the iron to form rust (Fe_2O_3). As the oxygen leaves the air, water will rise in the cylinder to take its place, volume for volume.

At the next laboratory period, adjust the water levels inside and outside the cylinder until they are equal. (Either raise the cylinder or add water to the beaker.) Do not permit any of the gas remaining in the cylinder to escape. When the levels are equal, move the rubber band to the water level. This will mark the volume of residual gas when it exerts a pressure on the water inside the cylinder that equals the pressure exerted by air on the water outside the cylinder. (In other words, the entrapped gas is neither compressed nor expanded but is at atmospheric pressure.) Again record the temperature and barometric pressure.

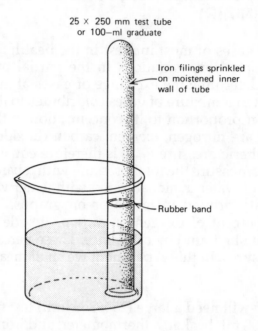

25 × 250 mm test tube or 100—ml graduate

Iron filings sprinkled on moistened inner wall of tube

Rubber band

Fig. 22. Removal of oxygen from air by reaction with iron filings.

Calculations. Record the data, and make the calculations called for on the Report Sheet as the experiment progresses in order to help your understanding of the very important concept of partial pressures.

You determined the original volume of air trapped by measuring the volume of the test tube filled to the brim. Determine the final volume of air remaining by measuring the water required to fill the tube or cylinder to the level of the thin rubber band.

P_1: You recorded the temperature when you started the experiment, and, by consulting a chart, you can find what the vapor pressure of water is at this temperature. This will be the partial pressure of water vapor, since we may reasonably assume that the entrapped air is saturated with water vapor. The partial pressure of the air without water vapor will be the difference between the vapor pressure of the water and the atmospheric pressure you recorded for the start of the experiment.

Part of the original gas volume and part of the residual gas volume are due to water vapor, however, and to get the *true volumes* of original air and residual air it is necessary to subtract the contributions to the volume made by the water vapor.

V_1: The volume of the "dry" air will then be found by multiplying the total volume of the wet air by the percent of the atmospheric pressure that is caused by the "dry" air. The same reasoning is following in correcting the volume of the unused air so that it represents the volume of "dry" air not used. The percent of oxygen in the air is then calculated from these corrected values of pressure and volume, by the difference between 100% and the percent of the residual air.

Sample calculations are summarized on the Report Sheet.

EXPERIMENT 11 PARTIAL PRESSURE OF OXYGEN IN THE AIR

Measurements at the Start of the Experiment:

Date _____ ; Barometric pressure (P_T)_____ mm Hg

Temperature _____ °C = _____ K (This is T_1)

Vapor pressure of water (P_{water})_____ mm Hg (from chart in textbook)

Pressure of the dry gas (P_T - P_{water}) _____ mm Hg (This is P_1)

Volume of air in the test tube _____ mL;

(P_1 x vol. test tube) ÷ P_T = Volume of dry air _____ mL (This is V_1)

Using the relationship: (P_1 x V_1) ÷ T_1 = (P_2 x V_2) ÷ T_2

where P_1, V_1, and T_1 are at laboratory conditions, and P_2, V_2, and T_2 are at Standard Conditions,

V_2 = (P_1 x V_1 x T_2) ÷ (T_1 x P_2) = Volume corrected to standard conditions

V_2 = _____ mL

Measurements at the End of the Experiment:

Date _____ ; Barometric pressure _____ mm Hg

Temperature _____ °C = _____K (T_1)

Vapor pressure of water _____ mm Hg

Pressure of the dry residual gas _____ mm Hg (P_1)

Volume of air in the test tube _____ mL

Volume of dry residual air _____ mL (V_1)

Volume of dry residual air corrected to STP _____ mL (V_2)

V_2 (original gas) - V_2(residual gas) = volume of gas used = volume O_2 =

_____ mL O_2 (continues on next page)

(Volume of Oxygen ÷ Volume of original gas) x 100 = percent O_2 in air =

_____ %

Calculate the % error = (___ % from experimental ÷ 21%) x 100 = _____ %

EXPERIMENT 12 FACTORS AFFECTING SOLUBILITY (HOW MUCH SOLUTE CAN BE DISSOLVED IN A GIVEN AMOUNT OF SOLVENT)

12 A. The Nature of the Solvent

Procedure. Place 2 mL of water in one test tube; 2 mL of hexane in a second; and 2 mL alcohol in a third. Be sure the second and third tubes are dry before adding the solvent. Into each test tube introduce a very small iodine crystal—try to have the crystals as nearly the same size as possible. Swirl the contents of each tube to encourage solution and <u>record what you observe </u>on the Report Sheet. Discard waste solutions in the designated waste receivers.

12 B. The Nature of the Solute

Procedure. Into three clean dry test tubes place 3 mL of hexane. Do not inhale hexane or get it on your skin. Use stoppers, not your finger or thumb, when shaking the test tube. Try to dissolve about 0.1 g sodium chloride, 0.1 g sucrose, and 1 drop of vegetable oil (or a very small shaving of paraffin) in the separate tubes of hexane. Stopper and agitate each tube and <u>record your observations on the Report Sheet.</u>

Repeat the experiment using 3 mL of water instead of the hexane. Make sure the stoppers are clean and dry. Do not try to use too much solute—adding solid beyond the true solubility may give a wrong report of "insoluble." Also, it is very important to make a physical effort to make the solute dissolve—stopper and agitate as before. Discard waste solutions in the designated receivers.

Repeat the experiment this time using 3 mL of alcohol.

12 C. Temperature

Procedure. Mix together 1 g of potassium nitrate and 2 mL of water in a clean test tube. Does the solid dissolve completely at room temperature? Warm the mixture gently, preferably using a boiling water bath, as you stir it with a glass rod. Record what you see. Then cool the mixture by holding the tube in a beaker of cold water. <u>Record and explain what you see</u>. If recrystallization does not occur spontaneously, the addition of a minute crystal of potassium nitrate may be necessary. Discard wastes in the drain in the sink.

Name ______________________________ Partner ____________________________

Section __________ Date ___________ Due Date ____________ Score _______________

EXPERIMENT 12 FACTORS AFFECTING SOLUBILITY

12 A. The Nature of the Solvent

1. Compare the solubility of iodine in the three solvents (state the polarity of each solvent).

In water:

In hexane:

In alcohol:

2. In which solvent did the color of iodine resemble its color in tincture of iodine?

12 B. The Nature of the Solute

3. In which solvent did the following dissolve most readily?

Sodium chloride ____________________________________

Sucrose ____

Vegetable oil (or paraffin) ___________________________

4. On the basis that water is very polar and hexane is nonpolar, classify the following *solutes* as polar or nonpolar.

Sodium chloride_______________ Sucrose _______________

Vegetable oil (or paraffin) ______________________

5. Explain using diagrams, the solubility behaviors seen in this part.

(Questions continue on the back of the page.)

12 C. Temperature

    6. Observations: at room temperature (did some, or all, dissolve?):

Observed when the mixture was warmed:

Observed when the solution was cooled:

    7. Explain what you saw in this part.

    8. If a crystal was added, what happened? If you obtained a supersaturated solution at any point, state how you recognized the fact and how you changed it into a saturated solution:

EXPERIMENT 13 FACTORS AFFECTING THE RATE OF DISSOLUTION AND RE-PRECIPITATION: KINETIC MOLECULAR THEORY

Experiment 12 dealt with the factors that affect solubility, or the amount of solute that can be dissolved in a given amount of solvent. The rate of solution is the speed with which the dissolution takes place and depends on the opportunities for contact between solvent and solute. As with a reaction, collisions are necessary for the solvent molecules to be able to separate the solute molecules from each other and disperse them singly into the continuous phase of the solvent. The frequency of collisions can be increased by making the solvent molecules move faster by stirring (or shaking) the mixture. Increasing the temperature or the surface area of the solute per unit of mass of the solute also increases the frequency of collisions. The solute for this experiment (parts A-C, and E) is a salt with a colored ion, which makes its movement into solution and its relative concentration in solution easy to follow.

Crystallization, or re-precipitation, also relies on collisions. This time it is collisions between the dissolved particles (ions) and the growing face of the crystal. The rate at which this is permitted to occur affects the overall orderliness of the process, as we will see.

13 A. Effect of Temperature

Procedure. Prepare two test tubes half full of distilled water, and heat one until the water boils. (Be sure to follow all of the precautions given as part of the "Techniques" section in Chapter 1 about boiling a liquid in a test tube.) Add to each approximately equal-sized crystals of the colored salt furnished for the experiment; observe the relative rates of dissolution and record on the Report Sheet.

13 B. Effect of Stirring or Shaking

Procedure. Prepare two test tubes half full of distilled water, and add to each approximately equal-sized crystals of the colored salt. Cork and shake one tube vigorously but let the other tube remain undisturbed. Again compare the rates of dissolution and record on the Report Sheet.

13 C. Effect of Particle Size

Procedure. In separate test tubes half full of distilled water place a large crystal of the colored salt and an approximately equal mass of the salt in powder form. Stopper and shake the two tubes vigorously and note the relative rates of dissolution. Record on the Report Sheet.

13 D. Supersaturation and Rapid Reprecipitation

Photographer's hypo is $Na_2S_2O_3 \cdot 5H_2O$. It is thus a hydrate or pentahydrate of sodium thiosulfate. If it is warmed to 45° - 50° C, it will "melt"; that is, the crystals will break up, and the liberated water of hydration dissolves the sodium thiosulfate, resulting in a solution. If this solution is protected from dust and mechanical jarring and allowed to cool to room temperature, a supersaturated solution of sodium thiosulfate will form.

Procedure. Obtain 2-3 g of crystals of photographer's hypo in a clean, dry, lint- and dirt-free test tube. Heat the tube cautiously and gently until the crystals "melt." Hold the tube well above the flame. Set the tube in the test tube rack and invert over it a small beaker to prevent dust from entering.

If the tube cools to room temperature without crystallization of its contents, seek out two neighbors who have obtained the same results. Let one person insert a glass rod and scratch the inner wall of one of the test tubes below the surface of the solution. Another person should drop a tiny crystal of "hypo" into the solution in another of the test tubes. The third may simply agitate the solution vigorously. Record your observations on the Report Sheet.

If you do the next part, watch for the difference in the sizes of crystals grown slowly (B) versus grown rapidly (A). Discard wastes as directed.

13 E. Growing Crystals Slowly (Group or Class Project)

This is a project in which three or four students could work as a team.

Procedure. Using detergent solution, thoroughly clean a 250-mL beaker. Rinse it with deionized (or distilled) water. Try to make the inside of the beaker completely dust-free. Transfer 43 g of copper sulfate pentahydrate ($CuSO_4 \cdot 5H_2O$) to the beaker and with a dust-free graduated cylinder transfer 50 mL deionized (or distilled) water to the beaker also. Place a watch glass over the beaker and warm the beaker until the copper sulfate dissolves. (You may stir the contents from time to time.) Set the beaker aside to cool to room temperature. Let the watch glass remain in place until the beaker has cooled until it is barely warm to the touch. Then replace the watch glass with a plastic film (e.g., Saran Wrap, Handi-Wrap), and set the beaker in your laboratory desk or other designated place to remain until the next laboratory period.

Enjoy the beauty of a mass of well-formed crystals of copper sulfate pentahydrate. It is impossible to imagine how the sharp planes and the well-defined angles of crystals could emerge unless the particles making up crystals (ions and molecules and, sometimes, atoms) stack together in very precise patterns. *Do not touch the blue crystals or solution with fingers.*

EXPERIMENT 13 FACTORS AFFECTING THE RATE OF DISSOLUTION AND RE-PRECIPITATION

Observations

13 A. Effect of temperature:

13 B. Effect of stirring:

13 C. Effect of particle size:

13 D. Supersaturation and Rapid Reprecipitation

13 E. Growing Crystals Slowly (Group or Class Project)

Questions

1. Explain the results seen in parts A-C using the Kinetic-Molecular Theory.

2. Compare the sizes and regularities of crystals grown slowly (E) with those forced to grow rapidly (D). (When crystals are grown slowly, the system is never far from being at equilibrium between the crystals and the dissolved solid.)

(Questions continue on the back of this page.)

3. In parts D and E, after the crystals start to grow, what would you have to do to stop growth? (Describe two or three things that would have to be done experimentally to stop the crystals from growing and to establish a new equilibrium.)

Warning: Since copper sulfate is a dangerous poison, you should not take its crystals from the laboratory. To loosen the crystals from the bottom of the beaker, gently poke at the edge of the mass with a metal spatula until parts of the mass start to break away from the glass. The rest will then loosen easily. Place the crystals and the blue liquid above them in the designated receptacle at the side shelf.

CHAPTER 6

Water, Solutions, and Colloids

EXPERIMENT 14 REACTION TO STRESS BY SYSTEMS IN DYNAMIC EQUILIBRIUM: CONCENTRATION EFFECTS

In a saturated solution of sodium chloride that is in contact with undissolved salt, solid sodium chloride is in equilibrium with dissolved sodium ions and chloride ions:

$$NaCl_{(s)} \rightleftharpoons Na^+_{(aq)} + Cl^-_{(aq)}$$

At any given temperature, the solubility of sodium chloride is a constant, and the product of the concentrations of the two ions is also a constant, called the solubility product constant and symbolized as Ksp. If anything happens to increase the concentration of one of the ions, the constancy of the ion product demands that the concentration of the other ion must decrease—which means that some sodium and chloride ions must combine and separate from the solution as solid sodium chloride. Thus, if either Na^+ or Cl^- is added to the saturated solution, the equilibrium will shift to the left. The "stress" is the added Na^+ or Cl^-. By shifting to the left and taking solid NaCl out of solution, the stress is absorbed.

Procedure. 14 A. Place 2 or 3 mL of saturated sodium chloride solution in a test tube and add 1 (one) crystal of NaCl. If it dissolves (was the solution saturated?), add another until at least one crystal of NaCl can be seen in the solution. Now add concentrated hydrochloric acid drop by drop until additional crystals separate from the solution.

14 B. Repeat A, using saturated ammonium chloride instead of sodium chloride solution.

14 C. Aqueous ammonia is an equilibrium between ammonia gas dissolved in water and the ions NH_4^+ and OH^-:

$$NH_3 + H_2O \rightleftharpoons NH_4^+ + OH^-$$

Prepare a stock solution by adding to 25 mL of distilled water, 1 drop of concentrated aqueous ammonia and 1 drop of phenolphthalein solution. The phenolphthalein should give a pink color to the solution. (Why?) Carefully note whether you can smell the ammonia.

Use 4 separate 5-mL portions of this stock solution as follows:
  1. To 5 mL of this stock solution, add a very small amount of saturated ammonium chloride solution; record your observations.

2. To 5 mL of the stock solution, add dilute HCl drop by drop until the pink color is dispelled. Count the drops needed. Again record your observations.

3. To 5 mL of the stock solution, add a few drops of dilute sodium hydroxide solution. Record your observations, including any change in odor or color.

4. Heat 5 mL of the stock solution to boiling, carefully, for a minute or two. Cautiously note the odor of the hot solution. Cool the solution in a beaker of cold water or under the tap. Is the color different? Add drops of dilute HCl until all the color is gone, again counting the drops. Record your observations.

EXPERIMENT 14 REACTION TO STRESS BY SYSTEMS IN DYNAMIC EQUILIBRIUM

14 A. 1. Sodium chloride solution: What is the evidence for a shift in equilibrium?

2. In which direction did the equilibrium shift—toward products or reactants?

3. What ion caused the equilibrium to shift?_____

4. What ion was removed from solution to relieve the stress? _____

5. If solid sodium hydroxide were added to neutralize the hydrochloric acid, would this reverse the reaction and cause the precipitated sodium chloride to redissolve? <u>Explain</u>—write the equation for the equilibrium where $[Na^+]$ = $[Cl^-]$ in molar concentration, and then where Cl^- has been added in excess. Then show what happens if added Na^+ must be accommodated.

14 B. Ammonium chloride solution: 1. What is the evidence for a shift in equilibrium?

2. In which direction did the equilibrium shift?

3. What ion caused the equilibrium to shift? _____

14 C. 1. Ammonia solution plus saturated ammonium chloride solution: What common ion is added? _____ What do you observe? (Change in color? Odor? Precipitate? etc.)

EXPLAIN. (Which way would the NH_4^+ ion shift the equilibrium?)

2. Ammonia solution plus dilute HCl; number of drops added: _____.

Change in odor? _____ With what substance in the reversible equation would an ion from HCl react?

What way would this shift the equilibrium?_____

The *ions of a salt* are not volatile—a substance must be volatile in order to travel through the air to reach your nose. On this basis, explain your observations in (2)

3. Ammonia solution plus sodium hydroxide solution:

What common ion is added? _____ What did you observe?

How did this shift the equilibrium?_____

Explain your observations.

4. Ammonia solution heated to boiling; observations:

Difference in color?_____ Difference in odor?_____

Number of drops of HCl added: _____ (Compare with (2))

Remembering the gas laws, compare the gas tensions (partial pressure) of ammonia over an ammonia solution at room temperature and at 100 °C. If an ammonia solution were boiled, would the concentration of dissolved ammonia be affected? _____ Explain:

How would you account for the fact that a bottle of household ammonia loses its strength, even when "stoppered," if it stands on the shelf for a fairly long period? (Compare the partial pressure of the ammonia in the bottle with the usual partial pressure of the ammonia in room air.)

EXPERIMENT 15 SURFACE TENSION AND SURFACTANTS

Surface tension of water is caused by the attraction of the water molecules for each other; virtually no attraction exists between the surface water molecules and the air molecules just above them. At the interface between water and glass, some attraction between water and glass tends to pull the water molecules up along the surface of the glass. This creates the meniscus defined and illustrated in Fig. 14 in the discussion of volumetric measurements.

If water is dropped on a nonpolar surface, such as waxed paper, the water will tend to contract into rounded droplets. In this form the water presents as little surface contact as possible with the wax. This tendency to form a tight surface "skin" can be demonstrated by the floating of a needle on water in a teacup by "parlor magicians" or in the laboratory by the floating of sulfur powder on water in a test tube. The destruction of this "skin," or tension, by a surface-active agent can also be demonstrated by the parlor magician who floats a soap flake on the surface near the needle and allows it to dissolve. This reduces or destroys the attraction of the surface water molecules for each other. The floating needle is no longer supported and it sinks. In the laboratory when surface-active soap molecules or synthetic detergent molecules are introduced into the water before the powdered sulfur is dropped onto the surface, the powder no longer floats.

Procedure. Rinse four test tubes thoroughly so that traces of detergent are completely removed. In separate tubes, place a few milliliters of each of the following liquids: distilled water, 0.1% sodium chloride, 0.1% Ivory soap, 0.1% syndet (or 0.1% bile salt). Set the tubes side by side in a test tube rack and gently dust very small amounts of powdered sulfur onto the surfaces of each liquid. Tap the sides of the tubes gently and observe in which tubes sulfur most easily breaks through the surface of the liquid. Record your observations on the Report Sheet

EXPERIMENT 15 SURFACE TENSION; SURFACTANTS

1. Is the powdered sulfur a polar or a non-polar substance?_____

2. Sulfur broke through the water surface in the tubes containing:

3. Compare sodium chloride with sodium stearate (Ivory soap), synthetic detergent or bile salt in effect on surface tension:

4. Draw a diagram showing how detergents make oils soluble, emphasizing the polar and nonpolar parts of the surface-active agent, the polarity of water, and the nonpolarity of the oil and grease (see your text).

Consult your text on the role of soaps and synthetic detergents in making small droplets out of grease and oils on dishes, clothes, and skin and thereby dispersing the oil and grease and adhering soil in the wash water before sending them "down the drain." See your text also for the role during digestion of a natural detergent, bile salt, in emulsifying water-insoluble fats and oils from foods. This action helps expose dietary fats and oils to the digestive enzymes and aids in their absorption from the digestive tract into the circulatory system.

EXPERIMENT 16 COLLOIDAL DISPERSIONS VERSUS TRUE SOLUTIONS

True solutions, colloidal dispersions, and suspensions are all examples of matter dispersed in a liquid medium. The difference lies in the size of the dispersed particles. In a true solution, the dissolved particles are so small they do not settle, and they do not reflect and scatter light (the Tyndall effect). Colloidal particles also do not settle from solution, but their particles are much larger than those in true solution and do reflect light. They are usually kept in suspension by having some kind of hydrophilic ("water-loving") coating, such as soap or detergent, or they may bear like-charges of ions that repel each other and keep the particles from coalescing. With starch, as in this experiment, we have hydrated particles. Suspensions, on the other hand, are affected by gravity and do settle, the finer particles merely taking a longer time.

In this experiment you will compare certain properties of a colloidal starch dispersion with those of a calcium chloride solution. You will also study the Tyndall effect and the use of the iodine test for starch. Finally, you will prepare a nonaqueous solid-in-liquid dispersion, or gel.

16 A. Preparation of a Colloidal Dispersion of Starch in Water

Procedure. Place 1 g of starch in a 250-mL beaker. Carefully stir in 2 or 3 mL of water, leaving no lumps but a smooth slurry. Then, gradually and carefully add water to a total of 100 mL. Heat gently while stirring until the solution is near the boiling point and until a translucent solution results (2 or 3 minutes). Your preparation is now a 1% starch dispersion. Cool this dispersion to room temperature by swirling the flask in a stream of cold water.

Prepare dispersions of 0.1%, 0.01%, 0.001%, and 0.0001% starch from this 1% stock in the following manner: For the 0.1% dispersion, take 10 mL of the 1% dispersion and mix it with 90 mL water in a beaker or erlenmeyer flask. The new starch preparation is 0.1% starch. Use 10 mL of this and dilute it to 100 mL with water to make it 0.01%. Proceed from this dispersion in like manner to prepare the other starch dispersions. Be as precise as you can—remember, each dilution depends on the care with which you prepared those preceding. You will not need the full amounts of these mixtures. Reserve about two-thirds of a test tube of each of the others in labeled test tubes.

For part B, these test tubes should be clean, inside and out. Prepare also a 1% calcium chloride solution by dissolving 1 g of calcium chloride in 100 mL of water.

On the Report Sheet, note which dispersions appear cloudy and which (if any) are substantially as clear as the calcium chloride solution.

16 B. Tyndall Effect

Procedure. Using a beaker or a test tube rack to transport test tubes containing the starch dispersions and the calcium chloride solution, take the test tubes to the Tyndall apparatus and study each for the appearance of the Tyndall effect. For comparison purposes, use a test tube of distilled water as a control. Observation must be made at right angles to the light beam. Which starch dispersions (if any) fail to give the Tyndall effect?

(A Tyndall apparatus is a strong but narrowly focused light source in a dark box or a darkened room set up at right angles to the test tube holding the solution. The light passing through the liquid in the tube can be seen as a beam of scattered light, much like light passing through a dusty room.)

16 C. Iodine Test for Starch

The iodine test solution (iodine reagent) is prepared in the proportions of 20 g of potassium iodide and 10 g of iodine, together dissolved in 1 liter of solution. This has been done for you. Iodine is only slightly soluble in water (Experiment 12), but when iodide ions are present, iodine molecules form a complex with them, as I_3^-. This is soluble and releases iodine easily to starch.

Procedure. The iodine test for starch consists of adding a drop of iodine reagent to 1 mL of the solution to be tested in a white spot plate. Test 1-mL volumes of each of the starch dispersions with one drop of iodine reagent. Add a drop of iodine reagent to 1 mL of distilled water for comparison purposes. A yellow solution in the spot plate indicates the absence of starch. If a few blue granules are observed, it is a very weak test, but positive. A concentrated starch solution may give almost a black test.

On the Report Sheet, note which dispersions (if any) failed to give the iodine test for starch. If the tests must be made in test tubes instead of on spot plates, be sure that identical test tubes as well as identical volumes of each solution are used, and view down through the tube from the top against a white background on the desk. Compare the intensity of the colors.

17 D. Preparation of a gel ("Canned Heat")

(This makes a good demonstration.)

In part A of this experiment, you prepared a colloidal dispersion of starch in water; this consisted of hydrated starch molecules, which are too large to be dispersed individually but do not settle out because of their association with the water. Colloidal particles can become "solvated" in such a way that their dispersion is no longer a liquid but maintains the shape of the container in which it was formed. You are probably familiar with making a gel from dehydrated gelatin and hot water. When it cools it is hydrated gelatin and consists of a network of protein molecules holding as much as 50 times its own mass of water as a gel, which maintains the shape of the container. If too much water is added, or if the gel is allowed to warm up, the system then becomes a liquid, or sol. (If starch solution is made sufficiently concentrated, it will also form a gel.)

In the above examples the dispersing medium or continuous phase, is water. Other gel systems can be made using nonaqueous media as a continuous phase. "Canned heat" is an example.

Procedure. Place 15 mL of denatured alcohol in a small (50-mL) beaker. As you swirl the alcohol, add all at once 2 mL of saturated calcium acetate solution. Simultaneously with this addition, cease swirling the liquid in the beaker. After about 1 minute turn the beaker upside down (the material should no longer be fluid).

Cut a small piece of the gel (about 1 cm^3), place it on your wire gauze, and ignite it. The alcohol in the gel burns readily, giving heat just as an alcohol lamp does. The acetate ion ($CH_3CO_2^-$) also burns, but not so completely, leaving a slight amount of carbon on the white calcium oxide ash on the supporting wire.

EXPERIMENT 16 COLLOIDAL DISPERSION VERSUS TRUE SOLUTION

Indicate the relative strengths of the tests (or effects) in each situation listed in the table below by using (0) for no test or effect, (+) for a very weak test, (+) for a weak test, and an extra + for each successively stronger test (or effect).

		16 A Cloudy appearance	16 B Tyndall effect	16 C Iodine test
Water				
CaCl$_2$	1.0 %			
Starch	1.0 %			
Starch	0.1 %			
Starch	0.01 %			
Starch	0.001 %			
Starch	0.0001 %			

Which test, the Tyndall effect or the iodine test, is more sensitive for detecting starch in very dilute dispersions?

EXPERIMENT 17 HEMOLYSIS, CRENATION, AND DIALYSIS

17 A. Hemolysis and Crenation

When a drop of blood falls into tap water, a red color diffuses through the water since so much water diffuses into the cells they rupture (hemolyze). A drop of blood falling into a concentrated salt or sugar solution forms dark red globules or clumps because water diffuses from the cells, shrinking (or crenating) them.

These changes can be demonstrated using an overhead projector with a low-power microscope. It can also be done as group experiments using test tubes containing the designated solutions into which blood is dropped. The tubes are then swirled to mix and set aside to settle. Settling can be hastened if centrifuge tubes instead of test tubes are used and centrifuging is done after the solutions have had about 15 minutes to act on the blood.

Procedure. Label seven test tubes or small centrifuge tubes with the solutions listed below. Into the appropriate tube put 5 mL of the solution. Clean the end of a finger with alcohol, and using a sterile lancet, pierce the finger so blood can be dropped into each tube (one large or two small drop s in each tube). If the blood falls on the wall of the tube, it can be usually drawn into the liquid by tilting the tube so the drop of blood is bathed with the liquid. A clean stopper can also be used in the tube, but do not shake to mix—use only a swirling motion. The following solutions are suggested: (a) distilled water, (b) 0.5% NaCl, (c) 0.9% NaCl (isotonic), (d) 5% NaCl, (e) 0.15 M glucose, (f) 0.3 M glucose, (g) 1 M glucose.

Observe whether, during settling or centrifuging, any color is transferred to the solution or whether it remains clear and colorless. If possible, remove some of the sedimented cells and observe them with a hand lens to see if they are swollen or shrunk. Record your observations on the Report Sheet

17 B. Dialysis

Cellophane tubing is a convenient dialyzing membrane. In this experiment it will be easy to see what it does and does not allow to pass through it, much the same as occurs in cell membranes.

Equipment. Cellophane tubing, a glass stirring rod, 250- or 400 mL beaker, funnel.

Procedure. Obtain a 15-cm length of cellophane tubing and moisten in distilled water. Either make clean holes with a paper punch at each end, or slip paper clips over the ends so that it may eventually be supported in a beaker of water as shown in the Fig. 23.

Mix together 10 mL of a 1% calcium chloride and 10 mL of a 1% starch dispersion. By means of a funnel, transfer the mixture to the cellophane tubing, being careful that none of the mixture contaminates the outside of the tubing. (If it does, rinse it off with distilled water.) Suspend the system in a beaker of distilled water for half an hour (see Fig. 23).

After about 30 minutes, test the water in the beaker for the presence of calcium ions, chloride ions, and starch: Add 2-3 drops of 1 M Na_2SO_4 to a mL of the water. A white precipitate indicates calcium ions (now as $CaSO_4$). To another mL of the water, add 2-3 drops of silver nitrate (1 M $AgNO_3$). A white precipitate (AgCl) shows the presence of chloride ions. Perform the iodine test for starch.On the basis of your tests, answer the questions on the Report Sheet.

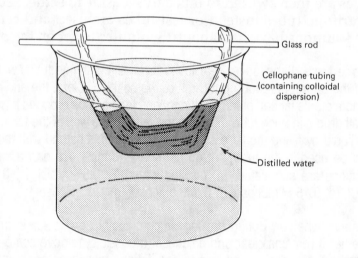

Fig. 23. Apparatus for dialysis

EXPERIMENT 17 HEMOLYSIS, CRENATION, AND DIALYSIS

17 A. Hemolysis and Crenation
Data and Observations:

Solution	Appearance After Centrifuging or Settling		
	Solution		Cells
	Color	Clarity	
(a) Water			
(b) 0.5% NaCl			
(c) 0.9% NaCl			
(d) 5% NaCl			
(e) 0.15 M glucose			
(f) 0.3 M glucose			
(g) 1 M glucose			

1. What happened to the blood cells in solutions less concentrated than 0.9% NaCl? (Did water flow into the cells, or out of the cells?)

2. What is a hypotonic solution?

3. What happened to blood cells in solutions more concentrated than 0.9% NaCl?

4. What is a hypertonic solution?

5. What is the normality of 0.9% NaCl? (Show calculations)

6. Assuming 100% dissociation, what is the effective molar concentration of ions in 0.9% NaCl?

7. Glucose does not dissociate. Which concentration of glucose was isotonic according to your tests?_____ Does this molar concentration correspond to the molar concentration of ions from NaCl assuming 100% dissociation?

17 B. Dialysis

1. Observation in starch test:

1.

2. Did any starch dialyze through the membrane?

3. Observation in test for chloride ions:

4.

5. Did any calcium ion dialyze through the membrane?

6. Observation in test for calcium ions:

7.

8. Did any chloride ion dialyze through the membrane?

9. Put (1), (2), (3) after these substances denoting the order in which they can most probably enter and leave cells:

large molecules (___), ions (___), small molecules (___).

CHAPTER 7   Acids, Bases, and Salts

EXPERIMENT 18 REACTIONS OF HYDRONIUM IONS

Obtain 40 mL of 3 M hydrochloric acid for this experiment.

18 A. Behavior of Aqueous Hydronium Ions With Metals

Procedure. Dilute 10 mL of 3.0 M hydrochloric acid to 20 mL to prepare a 1.5 M solution and divide it equally among four test tubes in the rack.

To the first tube, add a short length of copper wire; to the second, a clean iron nail or short length of iron wire; to the third, a small piece of mossy zinc; and to the fourth, a short strip of magnesium ribbon.

Observe which metal reacts most rapidly, which next most rapidly, and so on. If two do not seem to be reacting, gently warm the test tubes by placing them in a beaker of warm (e.g., 50 °C) water. (Do not boil the contents.) Warm them in the water bath until one metal appears to be reacting clearly more vigorously than the other. Discard left-over metal fragments in designated containers, not in the sink. Describe your observations on the Report Sheet.

Write net ionic equations for reactions that did occur. Arrange the metals in their order of reactivity. (This is the order of their tendency to become ions.)

18 B. Behavior of Aqueous Hydronium Ions With Metal Hydroxides, Carbonates, and Bicarbonates

Procedure. Place approximately 0.2 g of each of the following solids in separate test tubes:
    magnesium hydroxide
    sodium carbonate
    sodium bicarbonate

Carefully add 3 M hydrochloric acid in small portions (droppersful) to each tube until about 5 mL of acid has been added to each solid. On the Report Sheet, describe what you see. (Was heat evolved? Was a gas evolved? Did the solid dissolve?)

Write equations for the reactions on the Report Sheet.

Name ______________________________ Partner ______________________________

Section __________ Date ___________ Due Date _____________ Score ______________

EXPERIMENT 18 REACTIONS OF HYDRONIUM IONS

18 A. Behavior of Aqueous Hydronium Ions With Metals

Metal	Observations	Full ionic equation	Net ionic equation
Copper (Cu)			
Iron (Fe)			
Zinc (Zn)			
Magnesium (Mg)			

Order of reactivity of metals toward acid:

_______ $<$ ________ $<$ ________ $<$ _______

(Least active → most active.)

18 B. Behavior of Aqueous Hydronium Ions With Metal Hydroxides, Carbonates, and Bicarbonates

Substance	Observations
Magnesium hydroxide	
Sodium carbonate	
Sodium bicarbonate	

Equations

Magnesium hydroxide

Molecular equation

Full ionic equation

Net ionic equation

Sodium carbonate

Molecular equation

Full ionic equation

Net ionic equation

Sodium bicarbonate

Molecular equation

Full ionic equation

Net ionic equation

EXPERIMENT 19 PREDICTING REACTIONS USING SOLUBILITY RULES

Complete the pre-experimental work before coming to the laboratory.

Solubility Rules for Salts in Water

1. All lithium, sodium, potassium, and ammonium salts are soluble, regardless of the counterion (the oppositely charged ion).

2. All nitrates and acetates are soluble, regardless of the counterion.

3. All chlorides, bromides, and iodides are soluble, *except* when the counter ion is lead, silver, or mercury (I).

4. All sulfates are soluble, *except* those of lead, calcium, strontium, mercury(I), and barium.

5. All hydroxides and metal oxides are insoluble, *except* those of the Group 1A cations and those of calcium, strontium, and barium.

6. All phosphates, carbonates, sulfites, and sulfides are insoluble, *except* those of the Group 1A cations and NH_4^+.

By "solubility" we mean to the extent of a 3% to 5% solution. The rules do not cover all cases; there are exceptions. An exception might, in fact, be included in the experiment. Hence, make your best predictions based on the solubility rules as given, and observe very carefully what, in fact, does happen.

Assume the availability of separate solutions of each member of the pairs of compounds listed on the Report Sheet for Experiment 19. Consult the solubility rules listed above, and then study each pair and decide if, in solution, the two would react substantially with each other. If you predict that they would react, write an ionic equation for the reaction. Also write what you would expect to observe as evidence of a reaction (e.g., change in odor, evolution of a gas, separation of a precipitate, the evolution of heat). When you come to the laboratory, have your instructor check your work, and then test your predictions, recording your actual observations.

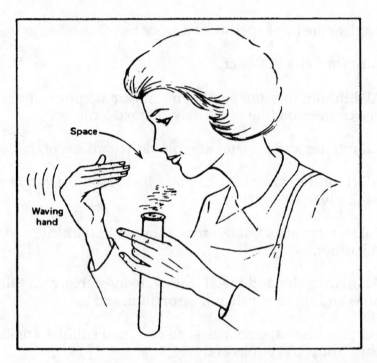

Fig. 24 Technique for testing odors. Waft vapors to the nose with your hand.

Procedure. Perform the tests by mixing small (1 mL) portions of the solutions of each pair. Be sure your test tubes are clean! Start by adding one drop of the second solution to note how rapidly a precipitate will form (how insoluble the product is). Does it take the full 1 mL of the second solution to cause the precipitate? If a gas is produced, smell it very cautiously. (See Fig. 24.)

EXPERIMENT 19 PREDICTING REACTIONS USING SOLUBILITY RULES

Predictions (Preexperimental Work):

Pairs of compounds to be combined:	If a reaction is expected, write the net ionic equation for the reaction:	Actual observations
$Ba(NO_3)_2 + Na_2SO_4$		
$Ca(NO_3)_2 + Na_2SO_4$		
$Ca(NO_3)_2 + NaCl$		
$Na_2CO_3 + Ca(NO_3)_2$		
$Na_2CO_3 + AgNO_3$		
$Na_2CO_3 + ZnCl_2$		
$Na_2SO_4 + ZnCl_2$		
$Na_2SO_4 + CuCl_2$		
$Na_2SO_4 + Pb(NO_3)_2$		
$Pb(NO_3)_2 + CuCl_2$		
$AgNO_3 + Na_2SO_4$		
$AgNO_3 + NaCl$		

Your Instructor may designate others. (Questions continue on back of page.)

1. Use this space to describe situations in which prediction did not match observation.

2. Write net ionic equations that describe any unexpected precipitates.

Schematic for reagent combinations:

	Na_2SO_4	Na_2CO_3	$Ca(NO_3)_2$	NaCl	$ZnCl_2$	$CuCl_2$	$Pb(NO_3)_2$	$AgNO_3$
$Ba(NO_3)_2$								
Na_2SO_4								
Na_2CO_3								
$Ca(NO_3)_2$								
NaCl								
$ZnCl_2$								
$CuCl_2$								
$Pb(NO_3)_2$								
$AgNO_3$								

Combinations taken from the very lightly shaded area may result in reactions, also, (and with the permission of your instructor may also be tried,) but some of these may not be simple precipitation reactions. They may result from acid-base or redox reactions whose products are solids. If you do try them, be prepared to write an equation for reactions observed.

EXPERIMENT 20 TESTS FOR VARIOUS IONS

A hospital laboratory report for blood or urine can list many inorganic substances, mostly in the form of ions. Some of these substances must be detected by special tests requiring more sophisticated equipment than we find in most college laboratories for beginning courses. Other ions as you have already discovered (Experiment 19), can be detected because they react with special ions to give a precipitate. Remember that a report may give units for "sodium," "potassium," or "calcium," but these are present as the ions. The above elements can never be present in the body because they react with water producing heat, hydrogen gas, and the hydroxide of the metal, none of which can be tolerated by the body.

It is important that both the solution to be tested and your test reagent solution are clear, not cloudy, when you start. Sometimes only a slight clouding of the solution after mixing the solutions is considered a positive test. A good example is the addition of a drop of silver nitrate to tap water, which probably contains a trace of chloride ion.

Equipment. Obtain several large test tubes. Clean them and rinse well with distilled water.

Solutions to be Tested. Each of the ions to be tested for is present in a separate solution. Find the desired ion in the formula on the label. For example,

the NH_4^+ ion needed will be found in the solution marked NH_4Cl, 0.1 M.

Tests for each ion:

ION TESTED FOR:	PROCEDURE:
NH_4^+ (0.1 M)	To 5 mL of the solution to be tested, add 1 mL 6 M NaOH and warm gently. A piece of moist litmus held at the mouth of the test tube (but not touching) should turn blue. Odor of ammonia may be detected.
Ba^{2+} (0.1 M)	To 2 mL of solution to be tested, add 1 mL of 0.1 M Na_2SO_4 solution. Look for a white precipitate.
Ca^{2+} (0.1 M)	(a) To 2 mL solution, add 1 mL 1 M Na_2CO_3 solution. Look for a white precipitate. (b) To a fresh 2 mL sample of solution, add 1 mL 4% ammonium oxalate, $(NH_4)_2C_2O_4$ solution. This precipitate may take several minutes to appear. Be patient!

Fe^{3+} (0.001 M)	(a) To 2 or 3 mL of solution, add 3 drops of concentrated HNO_3 and carefully heat to boiling. Cool. Add a few drops of 0.1 M potassium ferrocyanide solution. A rich blue ("prussian blue") color should appear. (b) Take a fresh 2-3 mL sample of the solution and treat it with the acid as above. Add a few drops of potassium thiocyanate solution this time. A rich red-orange color appears.
Cl^- (0.1 M)	To 2 mL of the solution to be tested, add 1 drop 5% $AgNO_3$ solution, then add 1 drop of conc. HNO_3. The white precipitate that remains is AgCl. If the precipitate dissolved when the HNO_3 was added, Cl^- was not present.
CO_3^{2-} (0.1 M)	(a) To 3 mL of solution to be tested add 1 mL 3 M HCl. The production of gas (CO_2) when acid is added to either a solid or solution containing CO_2^{2-} or HCO_3^- is a common identification procedure. It happens so quickly that it is often overlooked. In dilute solutions the effervescence may be difficult to see. (b) To a fresh 2 mL sample of solution add 1 mL of 0.1 M $CaCl_2$ solution. (Ca^{2+} does not precipitate HCO_3^-.)
SO_4^{2-} (0.1 M)	To 2 mL of the solution, add 1 mL of 0.1 M $BaCl_2$ solution, then add 5 drops of 3 M HCl. $BaSO_4$ is insoluble in HCl, but other barium compounds, like $BaCO_3$ or $Ba_3(PO_4)_2$, will dissolve.
PO_4^{3-} (0.1 M)	To 1 mL of solution, add 3 drops of conc. HNO_3 and warm in a hot water bath at 50-60 °C. Add 1 mL of molybdate reagent. Return the test tube to the warm water bath. If PO_4^{3-} was present, you should see a yellow precipitate.

Your instructor may ask you to test an unknown. Divide the sample of unknown solution into eleven 2 mL parts and test each as above. Include your results on the report sheet.

Name _____ Partner _____

Section _____ Date _____ Due Date _____ Score _____

EXPERIMENT 20 TESTS FOR VARIOUS IONS

Ion tested for:	RESULTS AND NET IONIC EQUATION:	Results for Unknown:
NH_4^+		
Ba^{2+}		
Ca^{2+}	(a)	
	(b)	
Fe^{3+}	(a) No equation required; color?	
	(b) No equation required; color?	
Cl^-		
CO_3^{2-}	(a)	
	(b)	
SO_4^{2-}		
PO_4^{3-}	No equation required; color?	

CHAPTER 8

Acidity: Detection, Control, Measurement

The concentration of hydrogen ions in biological fluids is of the greatest importance. A convenient way of describing acidity is given by *pH*. A change of one pH unit means a tenfold change in the concentration of the hydrogen ion. Variations of less than one pH unit can mean the difference between life and death. Means of determining the concentrations of both free hydrogen ion present, and total available (potential) hydrogen ion will be studied next. They form part of the basis for understanding the control mechanisms for maintaining constant hydrogen ion concentration in various parts of the body, concepts that are basic to physiological chemistry.

EXPERIMENT 21 DETERMINATION OF pH

pH is a shorthand way of expressing the concentration of hydrogen ions. It may be defined by one or the other of these two equivalent equations:

$$[H^+] = 1 \times 10^{-pH} \text{ or } pH = -\log[H^+]$$

where $[H^+]$ is the concentration of H^+ in moles per liter.

For rough values of pH, an indicator (or a series of indicators combined in one solution) can be used. *Indicators* are usually very weak acids that have a different color when changed to their salt forms (by being neutralized by a base). Litmus is a common indicator that is red below pH 4.5 but blue above pH 8.5. "Universal indicators" are mixtures of indicators that show a gradation in color in different pH ranges. These can be absorbed into paper and used like litmus paper, or they can be dissolved and added dropwise to the solution to be tested.

The most accurate and precise method of measuring pH is based on the conductivity of a solution between electrodes designed to react to the presence of hydrogen ions. The instrument is called a pH meter. With the more sensitive pH meters, the pH can be measured to decimal places much more precisely than would be possible with the use of indicators. Another advantage is that a pH meter can respond to a colored solution, whereas the color might interfere with seeing the color change in an indicator.

Your instructor will tell you how you will measure the pH—by indicator paper, indicator solution, or pH meter, or perhaps by both indicator and meter. Solutions you may want to test are: tap water, household ammonia, distilled water, household vinegar, 0.01 M acetic acid, 0.01 M HCl (gastric juice), soap solution, and your own saliva. In addition, the class may be invited to bring in personally interesting samples such as soft drinks, mouthwash, eyewash, shampoo, fresh and canned fruit juice, tea, coffee, cleaners of various types, urine, and the like. *Nothing that is oil-based such as ointments or oily emulsions should be used.*

One interesting test is that of the rain. If rain will fall within 24 hours of your lab session, collect a sample in an clean container with an air-tight cover. At the same time, your instructor should have a container of distilled water available that has been standing open to the air for several days. Compare the pH of these two. Do not use that distilled water sample for any of the other parts.

Warning: These solutions, if made in the lab ahead of time for you, MUST be made from freshly boiled and cooled distilled water. Solutions thus made MUST be tightly capped at that time to prevent their pH from changing and giving you false results.

Procedure for Indicator Paper. Place in marked positions on a paper towel strips of the indicator paper about 2 cm long. Carry the arrangement to the bottles of the solutions to be tested. Add a drop of a solution to a strip and compare the resulting color with the color code that comes with the pH paper.

Procedure for Universal Indicator Solution. Put one drop of the indicator solution in each depression of a clean, dry, spot plate. (Be sure it does not change color, indicating contamination of the spot plate.) Then add 0.5 mL of the solution to be tested. Compare with the color chart furnished with the solution. (Or put 1 drop of indicator and 1 mL of solution in the bottom of a clean, dry, test tube and hold against a white background.)

Procedure for the pH Meter. It is very important that the pH meter be used properly and that it has been standardized against a buffer solution close to the pH of the solution being tested. The instructor will give directions that will ensure your making accurate measurements without harm.

EXPERIMENT 21 DETERMINATION OF pH

Solution or substance	pH and color, using:		pH meter reading
	Indicator paper	Universal indicator	
tap water			
boiled, distilled water			
0.01 M HCl			
0.01 M acetic acid			
household vinegar			
saliva			
detergent solution			
soap solution			
household ammonia			
0.01 M ammonia			
0.01 M NaOH			

1. Both the hydrochloric acid and the acetic acid were 0.01 M. Was the pH the same for both? Explain.

(Questions continue on the back of the page.)

2. The concentrations of the NaOH and the NH_3 were also the same. Was the pH the same? Explain.

3. Was there a difference between the pH of the water allowed to stand open to the air for several days and the pH of fresh rainwater? If "yes," discuss why this may be so, and whether the difference indicates a serious situation.

(Note: both will test somewhat acidic because each has absorbed carbon dioxide from the air. This dissolves to produce a mildly acidic solution. This is normal. We are interested in the *difference* between the pH of these two solutions, however.)

4. Based on your results, which has a greater departure from the pH of a biological fluid such as saliva, the detergent or the soap?

EXPERIMENT 22 HYDROLYSIS OF IONS: HOW SALTS CAN CHANGE THE pH OF AQUEOUS SYSTEMS

Do the PREDICTIONS on the Report Sheet before coming to the laboratory.

Remember which kinds of ions interact with water to generate an excess of either H^+ or OH^- ions in the solution and so change the pH. The change in pH can be detected using a pH indicator like litmus or universal indicator.

For example, as sulfide ions enter solution (from the dissolution of solid Na_2S) they set up the following equilibrium.

$$S^{2-} + H_2O \rightleftharpoons HS^- + OH^-$$

The solution now has a (slightly) higher concentration of hydroxide ion than hydrogen ion, so the solution tests basic to universal indicator. The reaction of S^{2-} with water is an example of the *hydrolysis of an anion*. Notice that S^{2-} is the anion of a weak acid, HS^-. Such ions can be expected to hydrolyze like S^{2-}.

The behavior of an aluminum ion in water is an example of the *hydrolysis of a cation*. The Al^{3+} ion forms a hexahydrate, $Al(H_2O)_6^{3+}$, which is part of the following equilibrium.

$$Al(H_2O)_6^{3+} \rightleftharpoons [Al(H_2O)_5(OH)]^{2+} + H^+$$

Because of the H^+ generated in this equilibrium, the solution of $Al(H_2O)_6^{3+}$ tests acidic to universal indicator. Nearly all cations with charges of 2+ or 3+ behave similarly to the aluminum ion in water. The ammonium ion is a nojnmetal cation that also hydrolyzes to give an acidic solution.

$$NH_4^+ + H_2O \rightleftharpoons NH_3 + H^+$$

You will be asked to use universal indicator to test the solutions of several salts to see if they are acidic, basic, or neutral. *Before coming to the lab*, however, make a prediction for each of the salts listed in the first column of the Report Sheet. In the appropriate columns write the formulas of the cations and the anions. Is the anion that of a *weak acid*? If so, the anion can hydrolyze like the sulfide ion mentioned above. Is the cation *not* of a Group 1A element? If so, the cation can hydrolyze like the aluminum ion. Decide if H^+ or OH^- is predicted to be in excess in the solution. If a salt is from a weak acid and a strong base, which of its ions, either anion or cation, hydrolyze? Should the salt make the water acidic, basic or leave it neutral? Fill out your predictions on the Report Sheet. When you get to the laboratory, test your predictions by the following procedure, and record the results.

All solutions MUST be freshly prepared from BOILED distilled water!

Procedure. It is best to make these tests using a drop of universal indicator in a spot plate, adding about 0.5 mL of the solution to be tested and comparing with the color code for the indicator. The pH of the solutions can also be roughly determined as "acidic" or "basic" by adding a drop of the solution to a piece of litmus paper on a paper towel. Indicator paper may also be used, but it is found that unless the paper has a pH 7 color to start with (i.e., green), totally neutral salts such as NaCl will not change the color and will make the salt solution appear to be slightly above or below 7, depending on the indicator paper.

EXPERIMENT 22 HYDROLYSIS OF IONS

Predictions and Observations:

Formula of Salt	Ions from the salt:		Ion that can hydrolyze: (neither?)	Which ion, H^+ or OH^-, is made by hydrolysis? (neither?)	Solution therefore should be acidic, basic, or neutral?	OBSERVED pH
	Cation	Anion				
$NaHCO_3$						
Na_2CO_3						
$NaCl$						
Na_3PO_4						
Na_2HPO_4						
$NaCH_3CO_2$						
Na_2SO_4						
KBr						
$MgCl_2$						
NH_4Cl						
$CuSO_4$						
$FeSO_4$						
$Fe_2(SO_4)_3$						
$AlCl_3$						

(Questions are on the back of the page.)

1. List the ANIONS that are part of known *weak acids*: Under each, write the formula of the weak acid.

2. List the ANIONS that are part of known *strong acids*: Under each, write the formula of the strong acid.

3. Are any CATIONS part of known *weak bases*? Which cations are they? Under each, write the formula of the weak base.

4. Are any CATIONS part of Group 1A elements? Which cations are they? Write their formulas.

5. Which CATIONS bear charges of 2+ or 3+? List them.

6. In which of the above groups did you find ions that could alter the pH of water? _____

If you found exceptions within any groups, what were they? Suggest a possible explanation.

7. Choose one <u>anion</u> that underwent hydrolysis and write an equation of that ion with water that shows what happened to raise the pH of water.

8. Choose one <u>cation</u> that underwent hydrolysis and write an equation of that ion with water that shows what happened to lower the pH of water.

EXPERIMENT 23 TESTING A SUBSTANCE FOR BUFFERING ABILITY AND BUFFERING CAPACITY

Review Experiment 14 concerning Le Chatelier's principle and the effect of stress on systems in dynamic equilibrium.

The control of pH within narrow limits in the body is accomplished by means of *buffer systems*. These must be combinations of a weak acid and the negative ion (anion) provided by one of its salts, or a weak base and the positive ion (cation) provided by one of its salts. A particular buffer system is thus a pair of substances, one of which will combine with excess hydroxide ion *or* excess hydronium ion, so as to keep the pH constant. Each buffer system can stabilize pH over a characteristic range. It also has a specific capacity—it does not have the ability to neutralize unlimited quantities of added H^+ or OH^-. The more H^+ or OH^- the system can thus neutralize without change in the pH, the greater is its *buffering capacity*. We shall examine these characteristics of buffers.

The three major buffer systems in the blood are phosphate ($H_2PO_4^-$/HPO_4^{2-}), carbonate (CO_2/HCO_3^-), and proteins (which are both weak acids and weak bases). We shall examine both the buffering ability and buffering capacity of the first two. The study of proteins as buffers will be dealt with in the experiments on proteins.

Procedure. Test tubes to be used in this experiment must be clean. Use a small portion of distilled water as the final rinse. The effect on the pH of each buffer when a small amount of acid and of base is added will be compared with the effect on the pH of distilled water (an unbuffered medium) when acid or base is added in the same amount.

23 A. Unbuffered Solutions. Obtain about 40 mL of distilled water previously boiled to expel carbon dioxide and then cooled without mixing with room air. Prepare three test tubes containing 10 mL each of the special distilled water. Label them A-1, A-2, and A-3. Add to each the directed amount of universal indicator. If no directions are given, a rule-of-thumb is one drop of universal indicator per milliliter of solution.

Using a piece of glass tube with a constricted end (e.g., a pipet) blow into tube A-3 (the unbuffered water) for about one minute. Record what happens to the pH and write the equation to explain this on the Report Sheet. (This equation should be a permanent part of your vocabulary.) Does further blowing continue to change the pH? Set the A-group test tubes in the test tube rack until later.

23 B. Phosphate buffer. Prepare the phosphate buffer by mixing thoroughly 10 mL of 0.1 M Na_2HPO_4 with 10 mL of 0.1 M NaH_2PO_4. The buffer pair is $H_2PO_4^-$/HPO_4^{2-}. Add the directed amount of universal indicator. Divide the resulting solution into two equal volumes in test tubes labeled B-l and B-2. Record the pH as shown by the indicator.

23 C. Carbonate Buffer. Prepare the carbonate buffer by placing 20 mL of 0.1 M NaHCO$_3$ in a wide test tube (about 20 X 150 mm); add universal indicator as directed, and record the pH. Using a glass tube with a constricted end, blow into the solution as continuously as possible for at least two minutes, preferably longer. The original pH of the NaHCO$_3$ should be lowered at least one unit. The carbon dioxide from your breath will dissolve in the water in the solution to form CO$_2$(aq), giving you the buffer pair CO$_2$(aq) / HCO$_3^-$. Divide this solution into two equal volumes in test tubes labeled C-1 and C-2

23 D. Other solutions. Prepare 2 labeled test tubes containing each of the other solutions that your instructor has made available, if any. See the Report Sheet for suggestions. Set these aside and test them according to the next two parts that follow.

Demonstration of Buffering Ability:

Procedure. Set the test tubes labeled A-1, B - I, and C-1, side by side in the test tube rack. Add one drop only of 0.5 M HCl. Mix the contents of each tube well and record the new pH. If no change occurs, the solution has the ability to buffer. Test any other solutions given, such as (NH$_4$)$_2$CO$_3$, as well. Record your results on the Report Sheet. A buffer must also be able to neutralize base and keep the pH from changing. To each of the tubes just treated with acid, add one drop of 0.5 M NaOH. Again, mix well, and record the new pH.

Demonstration of Buffering Capacity:

Procedure. To test tubes B - I, C-1, and a sample of each of the solutions that showed buffering ability from the previous section, add 0.5 M NaOH dropwise, counting the drops, just until the blue-green color of the pH 9 region of the indicator is reached. Record the number of drops required to reach pH 9 in each case.

Now take test tubes labeled B-2, C-2, and a fresh sample of each type of other solution used above, and add 0.5 M HCl dropwise, counting the drops, until the orange of the pH 5 region of the indicator is reached. Record your results on the Report Sheet. If you notice a gas given off with any of the solutions, note this, and the pH at which it occurred.

Finally, still using tubes B-2, C-2, and the others to which HCl has just been added, add 0.5 M NaOH drop by drop, counting the drops, to return the solution to the original pH (see first part of Report Sheet). Are you surprised by the results? What has happened to the carbonate buffer? Did the same thing happen to the phosphate buffer? Compare the stability of the carbonate buffer to the stability of the phosphate buffer. Consider how the removal of carbon dioxide from the blood by the lungs could affect the pH of the blood (see Textbook).

Name _____ Partner _____

Section _____ Date _____ Due Date _____ Score _____

EXPERIMENT 23 BUFFER SYSTEMS

TESTING BUFFERING ABILITY:

Tube or solution	Original pH	After 1 drop of acid	After 1 drop of base	Conclusion?
A-1				
B-1				
C-1				
$(NH_4)_2CO_3$				
$NH_4CH_3CO_2$				
NaCl				

TESTING BUFFERING CAPACITY:

Solution	Tube	No. of drops of NaOH to reach pH 9	No. of drops of HCl to reach pH 5	No. of drops of NaOH to return to original pH
Unbuffered water	A-2			
$H_2PO_4^-/HPO_4^{2-}$	B-2			
$CO_2(aq)/HCO_3^-$	C-2			
$(NH_4)_2CO_3$				
$NH_4CH_3CO_2$				
NaCl				

Acidity: Detection, Control, Measurement

1. The pH of unbuffered water before blowing exhaled air into it_____

and after._____ Write the two-part reaction that explains this difference:

2. Which of the solutions you made or tested behaved as buffers? Give names of substances, not tube numbers.

3. Is the purpose of a buffer system to keep a solution neutral?_____ If not, what is the purpose?

4. Write net ionic equations to show how the phosphate buffer neutralized both acid and base.

5. Write net ionic equations to show how the carbon dioxide/bicarbonate ("carbonate") buffer neutralized the acid and the base.

6. Did the carbonate buffer lose any gas while HCl was added?_____ What was the gas? (Show a net ionic equation.)

7. Which solution showed the greatest buffering *capacity?*

8. Is there a connection between this loss of gas and the lower capacity of the carbonate buffer on the second addition of NaOH?

Explain (using net ionic equations).

EXPERIMENT 24 EFFECT OF GASES ON pH OF BUFFERED AND UNBUFFERED SOLUTIONS

In Experiment 23 you investigated the behavior of various buffer solutions toward a water solution of acid or base.

In this experiment, the added acid or base must enter the solution as a gas, by diffusion. Look for differences and similarities to the totally aqueous system. Also, the effect of these same gases on a diluted solution of the same buffer will be investigated. Will dilution affect buffering capacity?

(Partners are suggested)

24 A. Adding Carbon Dioxide.

Procedure. Put 5 mL of the following solutions into clean test tubes: distilled water, 1 M NaCl, phosphate buffer (as supplied), and phosphate buffer diluted one in three volumes of water. Into each put 4 or 5 drops of universal indicator. Record the pH on the Report Sheet. Using a clean glass tube, blow your breath slowly through each solution for one minute. What is the pH after this treatment? Does more blowing cause the pH of the buffered solution to change?

24 B. Hydrogen Chloride and Ammonia Vapors.

Procedure. Place seven clean test tubes of uniform size, marked 1 through 7, in a test tube rack. Using a pipet so as to keep the walls of the tubes dry, add 5 mL of distilled water to tubes 1 through 3.

Add 5 mL of the phosphate buffer provided to 10 mL of distilled water, mix well, and pipet 5 mL of this diluted buffer into each of two test tubes. Label these tubes 4 and 5.

Into two more test tubes, labeled 6 and 7, pipet 5 mL each of the undiluted buffer provided.

Therefore you should now have 3 tubes with distilled water, two with diluted buffer, and two with undiluted buffer. Add 5 drops of universal indicator to each of the seven test tubes. Record the color, and the pH, of each.

Obtain seven rubber stoppers that fit the tubes easily. Cut filter paper into seven strips (5 to 6 cm long), a little narrower than the test tube diameter. Hold the filter paper across the small end of the rubber stopper and add 3 drops of <u>ammonium chloride</u> to the paper; push the stopper into the tube marked 1. Push it in firmly, while the test tube is still in the rack, being careful not to disturb or agitate any of the others.

With the same care, add 3 drops of <u>concentrated HCl</u> to the paper on stoppers for test tubes 2, 4, and 6, and immediately stopper the tubes.

Repeat the procedure for test tubes 3, 5, and 7, this time using <u>concentrated ammonia</u> solution. (Continued on back of page.)

Leave the tubes completely undisturbed. Watch them, and record any changes in the solutions in the tubes.

Note the time required for changes to occur.

Discussion. The ammonium chloride is a "control." Vapors of the acid (HCl) and the base (ammonia) will encounter the surface of the solutions by diffusion, and then depend on solubility in the water to enter it. Kinetic theory tells us that the random motions of molecules in the solution will eventually carry the solute particles throughout.

Dispose of the strips of paper in containers in the hood, not the wastebasket. (Why?)

EXPERIMENT 24 EFFECT OF GASES ON BUFFERED AND UNBUFFERED SOLUTIONS

24 A.	Water	NaCl	Diluted buffer	Original buffer
pH of Solution at Start:				
pH of Solution after 1 Minute of Breath Mixing:				

1. Did continued blowing cause a change in pH of the buffered solutions? (If so, how long continued?)

2. Write a net ionic equation to show why the pH changed in the water and the NaCl solutions.

3. Write net ionic equations to show why the pH remained unchanged in the $H_2PO_4^-$/HPO_4^{2-} solutions.

24 B.

Tube No.	Solution	Solution above tube	Evidence of diffusion	Time at evidence of diffusion
1	Water	NH_4Cl		
2	Water	HCl		
3	Water	NH_3		
4	Diluted buffer	HCl		
5	Diluted buffer	NH_3		
6	Buffer	HCl		
7	Buffer	NH_3		

EXPERIMENT 25 TOTAL ACID CONTENT: ACID-BASE TITRATIONS

The purpose of these experiments is to acquaint you with one of the most common and the most important methods of analysis, volumetric analysis.

Procedures.

Your instructor will demonstrate proper techniques for cleaning and assembling and using the equipment.

Preparation and Cleaning of Equipment. Obtain from the supply area, depending on instructions given in the laboratory, either two 50-mL burets or one 50-mL buret and a volumetric transfer pipet (Fig. 16). The pipet could be a 20-mL or 25-mL type; your instructor will specify.

Wash the buret(s) and the pipet thoroughly and rinse them several times with tap water. Then rinse well with distilled water. You should similarly clean a 125 mL or 250-mL Erlenmeyer flask. (Note: If water forms beads on the inner walls of any of these items, they are not clean enough. Water does not bead on a thoroughly clean glass surface.)

How To Prepare a Buret or Pipet for Delivering a Solution. The cleaned and rinsed buret or pipet is filled with a small amount of the solution that will later be transferred from it. (Pipets should be filled with the aid of a suction bulb.) By "small amount" we mean about 5-mL for a buret and 3-5 mL for a pipet. It has to be enough to make contact with all of the inner wall of the buret or pipet as you tilt it nearly to the horizontal and slowly rotate. The objective is to "coat" the inside of the glass with the solution you are about to fill with. Do this rinsing operation three times. (Do not forget the buret tip.) The liquid now remaining in the pipet or buret is not rinse water from the initial washing. It is identical with the solution that you will place in the buret or pipet. Unless you prepare these volumetric items in this manner, the initial rinse water will dilute the solution you put in the buret or pipet and dilute it to an extent you do not know.

How to Use a Buret. Select a buret and be sure the stopcock is well greased and "seated." Rinse the buret and half fill it with water. Place one hand against the buret opposite the stopcock handle, and try working the stopcock. Press gently with the other hand so the stopcock does not work loose and leak. Practice releasing one drop of water at a time. Read the volume of the liquid in the buret before and after releasing one drop. Note that the numbers read from the top of the buret to the bottom, so you are reading the volume actually used. Note also that you should read the same part of the meniscus each time, and your eye should be in line with this level at the time of reading. See Techniques for reading volumetric glassware in Chapter 1 of this manual.

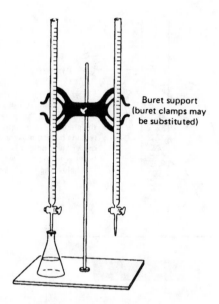

Buret support
(buret clamps may
be substituted)

Fig. 25 Titration assembly.

To save time in this Experiment, the concentration of the sodium hydroxide solution has already been determined. (We say that it has been *standardized*.) Be sure to record the molarity on your Report Sheet.

25 A. Standardization of Dilute Hydrochloric Acid. The proper technique for titration will be demonstrated by your instructor. Obtain approximately 120 mL of stock hydrochloric acid solution, and 120 mL of standard sodium hydroxide solution, each in separate, dry, labeled containers. You will determine the acid's molarity by titrating it against the standard NaOH solution provided.

Rinse out a buret with the standard sodium hydroxide solution three times using about 5 mL of solution each time. Then transfer 50-51 mL of solution to the buret. Drain out a few milliliters to rinse out the buret tip, working out any air bubbles in the tip as you do this. Mount the buret vertically on a ring stand with a buret clamp. (Fig. 25).

The unknown hydrochloric acid solution may be accurately measured and transferred either with a second buret or with a volumetric transfer pipet. If you use a second buret, rinse it with the hydrochloric acid solution three times by using about 5 mL each time. Then fill the buret to just above the zero line and drain out a few milliliters of the acid to rinse and fill the buret tip. This buret is also mounted on the ring stand. (Fig. 25). Touch off any droplets adhering to the tip. Allow a minute for the buret walls to drain and to detect any leaking at the stopcock. Read and record the exact initial volume, to the second decimal place, of both burets. Many students benefit from a verification of this reading by the instructor or another student. Accuracy in reading the buret volumes is important (Fig. 26).

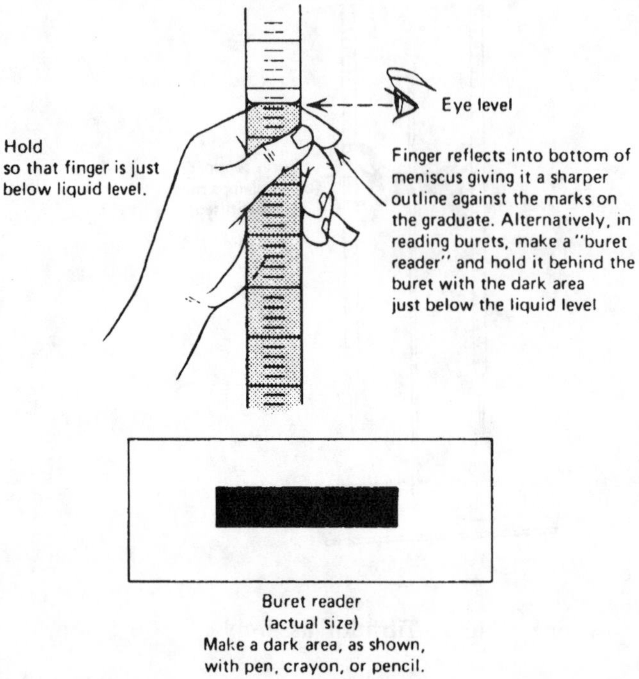

Hold
so that finger is just
below liquid level.

Eye level

Finger reflects into bottom of
meniscus giving it a sharper
outline against the marks on
the graduate. Alternatively, in
reading burets, make a "buret
reader" and hold it behind the
buret with the dark area
just below the liquid level

Buret reader
(actual size)
Make a dark area, as shown,
with pen, crayon, or pencil.

Fig. 26 The meniscus becomes more sharply outlined if an object darker than the liquid is made to reflect
into the meniscus. Read to the tenth's place. Estimate to the even hundreths place.

If you use a pipet, transfer its stated capacity of hydrochloric acid solution to a clean 125 mL or
250 mL Erlenmeyer flask. If you use the second buret, drain out about 20 mL of the acid solution
into the Erlenmeyer flask (record the exact volume), touching off any droplets to the inside of the
flask. With a little water from a wash bottle or a medicine dropper, rinse any solution clinging to
the glass down into the solution in the flask. Add two drops of phenolphthalein indicator solution
to the contents of the flask.

Put the flask with the acid solution under the buret and slowly and carefully add sodium
hydroxide to the acid, about a milliliter at a time. Between additions, swirl the flask to mix the
contents. When you begin to see a pink color lasting more than a second before you swirl the
flask, start adding the sodium hydroxide in smaller amounts, and swirl the flask each time. You
are nearing the endpoint.

Continue to add sodium hydroxide, drop by drop, swirling the flask after each addition, until a
faint pink (not a shocking pink) color persists for at least 15 seconds. Near the end, try adding
fractions of drops, touching each fraction to the inner wall of the flask and rinsing it down with a
small amount of water. A drop should make the difference between clear-and-colorless and a
clear-and-pink color which persists 15 to 30 seconds. Record the new volume.

If you overshoot the endpoint you will discover the advantage of using two burets in a titration. If
a second buret holds your acid, just let a few drops run out into the Erlenmeyer flask. That will

"erase" the deep pink color. Then reapproach the endpoint with sodium hydroxide, drop by drop. Record new volumes of both acid and base. If you used a pipet to add the acid, it is best to discard this result and start that titration over.

Discard the contents of the erlenmeyer flask, rinse it out well and place in it a new, carefully pipeted sample of acid. You may have to refill the buret holding the sodium hydroxide. It is best not to run out of solution in the buret in the middle of a titration. It means greater chances of error because of the greater number of readings you have to take if you refill the buret.

When you have a satisfactory endpoint, record final buret readings.

Repeat the titration with a new portion of hydrochloric acid in a clean flask. (Trials 2 and 3). From the data of each trial, compute the concentration of the hydrochloric acid solution. Calculate the average of the three. These should agree within 5%; if they do not, you may wish to do a fourth trial.

25 B. Analysis of Commercial Vinegar. Using a pipet, transfer 10.0 mL of the assigned vinegar to a clean 125 mL or 250 mL Erlenmeyer flask. (Remember, use a suction bulb to fill pipets!) Add about 20 mL of distilled or deionized water to the flask. This, of course, does not alter the number of moles of acetic acid in the flask. It simply dilutes the sample to make the titration a little easier. Add two or three drops of the phenolphthalein indicator.

Titrate the vinegar solution with standard sodium hydroxide solution as in Part A. Do three trials, recording all data on the Report Sheet. Calculate the concentration of the acetic acid from the equation:

$$\underline{M \text{ acetic acid}} \times 10.0 \text{ mL} = M \text{ NaOH} \times \text{vol. NaOH (mL)}$$

25 C. Determining the Effectiveness of Antacid Tablets By Titration. (Your instructor may invite you to bring your own antacid tablets for testing.) Weigh a piece of weighing paper to the nearest 0.01 g, then on it weigh an antacid tablet. Record both masses and the brand name of the antacid on the Report Sheet.

With a bulb and dry pipet, transfer 25.0 mL of the HCl solution used in Part A to a clean Erlenmeyer flask containing the antacid tablet. (You have already determined the concentration of the HCl solution.) Swirl the flask until the tablet dissolves. If it is necessary to use a stirring rod to help crush the tablet, be sure to rinse solution clinging to the rod back into the flask with 2 to 3 mL of water. The solution may not be clear because of starch or other binder added to hold the tablet together. Add five drops of phenolphthalein indicator. If a pink color develops, add 10.0 mL more HCl solution from a 10.0 mL pipet (dry or rinsed with the HCl solution). Keep careful record of the total HCl volume added. When the pink color of the indicator disappears, titrate the solution with the standardized NaOH solution from Part A until a permanent pale pink color is formed, as you did for the previous titrations.

Discussion. Your antacid tablet partially neutralized the HCl. The purpose of the titration was to determine how much of the known quantity of acid was not neutralized by the tablet. Therefore the number of moles of base added in the form of the NaOH solution represents that quantity. (This technique is called "back-titrating.")

$$M \text{ HCl} \times \text{vol. HCl (in L)} = \text{moles of acid (1)}$$

$$M \text{ NaOH} \times \text{vol. NaOH (in L)} = \text{moles of base } \underline{not} \text{ in tablet (2)}$$

Moles of base in tablet = (1) - (2)

Tabulate the class results on the blackboard, and compare the effectiveness of the various antacid preparations, <u>as moles of base per gram of tablet.</u> If there is a significant variation within one brand, check for sources of error—for example, reading the buret? weighing? Suggest some others.

EXPERIMENT 25 TOTAL ACID CONTENT: ACID-BASE TITRATIONS

25 A. Concentration of standard NaOH_____

	TRIAL 1	TRIAL 2	TRIAL 3
Vol. HCl used:			
Final buret read-ing (NaOH)			
Initial buret read-ing (NaOH)			
Vol. NaOH used, in liters			
Moles NaOH used (M x V)			

Moles of acid = moles of base at the equivalence point. Therefore,

Mol of HCl used:			
Calculated [HCl]:			
Average of the three trials =			

25 B. Vinegar number (or brand name) _____

Vol. vinegar used:			
Final buret read-ing (NaOH)			
Initial buret read-ing (NaOH)			
Vol. NaOH used, in liters			
Moles NaOH used (M x V)			

	TRIAL 1	TRIAL 2	TRIAL 3
Moles of acetic acid in each sample			
Calculated [acetic acid]			
Average of the three trials =			
Converting this to a mass/volume % =			

If more than one brand of vinegar was used by different groups, compare and comment:

25 C. Antacid brand_____

	TRIAL 1	TRIAL 2	TRIAL 3
Mass of tablet (g)			
Total volume HCl			
Moles acid used (1)			
Final buret reading (NaOH)			
Initial buret reading (NaOH)			
Moles NaOH used (2)			
Moles of base in antacid tablet (1) - (2)			
Moles base/gram of tablet			
Average of the three trials =			

Cost per tablet, this brand:_____ Cost calculated as moles of base/penny ($0.01)_____

CHAPTER 9 | Introduction to Organic Chemistry

EXPERIMENT 26 IONIC AND MOLECULAR COMPOUNDS

The type of bonding in a compound can greatly influence its chemical and physical properties. For example, ionic bonds allow such properties as solubility in water and electrical conductivity. Covalent (molecular) compounds are more likely to be insoluble in water (although this is not always true by any means!), and many will burn or have a low melting point.

The shape and composition of a molecular compound contributes to its solubility properties as well. Once certain principles are learned, one can predict water solubility or insolubility from a molecule's shape and composition (and therefore its polarity).

Ionic compounds are more likely to have significantly higher melting points than those of molecular (covalent) compounds, and are solids at room temperature. In contrast to this, can you think of any molecular compounds that are liquids, solids, or gases at room temperature?

Your substances to test may come from the following list, or your instructor may provide others:

potassium chloride	sucrose
methyl alcohol	benzoic acid
sodium acetate	calcium nitrate
toluene	isopropyl alcohol
potassium dichromate	nickel(II) chloride

26 A. Procedure. Descriptive. Obtain about 5 mL (if a liquid) or 2 g (if a solid) of each of the substances provided. On your Report Sheet, write down the name, chemical formula, and a short description of physical properties. (Do not include taste!)

26 B. Solubility tests: We use the generalization "like dissolves like" to describe the solubility properties of polar and non-polar substances. Water is used as the polar solvent here, and hexane as the non-polar solvent. (Hexane should not be touched, must be used only with good ventilation, and must be disposed of in a proper container—not down the sink drain. Ask your instructor what should be done with it.)

All test tubes must be clean and <u>dry</u>. Obtain 15-20 mL of hexane. Using fresh 2 mL portions of this solvent, test the solubility of each substance by adding 2-3 drops (or a pea-sized sample of a solid). Mix well with a clean, dry stirring rod. Record the results on your Report Sheet.

Now repeat the solubility tests for each substance, but this time use water as the solvent. Again record your results.

26 C. Ignition test: If you have been sloppy today about your safety glasses, make sure they are on your face now! Place 1 mL of liquid, or a pea-size sample of a solid, in a clean dry evaporating dish set on the wire gauze on a ring stand. Using a bunsen burner flame, cautiously test each substance individually for evidence of melting, then turn the flame directly on the substance very cautiously, and see whether it will burn, discolor, or char. Record your observations on the Report Sheet.

26 D. Conductivity: (This part may be done as a demonstration.) If you do this testing instead of having a demonstration, be sure to follow the Instructor's procedure for cleaning the electrodes between tests.

Warning: do not touch the electrodes with your hands or fingers at any time.

(For those substances that dissolved in water only:) Take a water solution of each soluble substance to a conductivity apparatus (usually a light bulb attached to a pair of electrodes and connected to a power source.) It is important that the electrodes be rinsed well with distilled water between each test to avoid obtaining confusing results.

Which ones cause current to flow through the solution? What does this mean? Record your observations on the Report Sheet.

EXPERIMENT 26 IONIC AND MOLECULAR COMPOUNDS

Name and Formula	Physical Description	Solubility in: water--hexane		Ignition Test Results	Conductivity Results

Use additional paper if necessary. Examine your results. On the next page you will be asked to decide whether each was an ionic or a molecular compound.

Questions:

1. Of all of the tests and examinations of the compounds on the list, which test would you say is the most reliable discriminator of ionic vs. molecular compounds?

2. Which of the substances you tested were definitely ionic compounds, and how did you decide?

3. Which of the substances you tested were definitely covalent (molecular) compounds, and how did you decide?

4. If there any for which there was an unclear choice, which were they? What traits gave you difficulty?

5. Choose one of the ionic compounds from question 2._____ By using diagrams, explain its solubility behavior in water:

6. Choose one of the molecular compounds from question 3. _____ Explain, using diagrams, its solubility properties by showing solute-solvent interactions (or lack of them):

8. Summarize the differences in physical behavior between ionic and molecular (covalent) compounds.

IONIC:

MOLECULAR:

EXPERIMENT 27 STUDIES IN STRUCTURAL ORGANIC CHEMISTRY: ISOMERISM

Use this exercise to help you mentally correlate a textbook diagram of a structural formula for a compound with its 3-dimensional ball and stick model.

Equipment. Obtain from the stock supply the following parts for "ball-and-stick" models:

Carbon balls, four holes	3
Hydrogen balls, one hole	8
Chlorine balls, one hole	2
Bonds, long sticks and short sticks* (or whatever "bonds" your model set provides)	4 and 8

*Ordinarily, short sticks are used for bonds from hydrogen and long sticks for all other single, covalent bonds; this makes the models more uniform for comparison purposes.

These parts are sufficient for the construction by one student of models up through C_3H_8 and $C_3H_6Cl_2$. For the larger models students may work in pairs, unless instructions for obtaining additional parts are given.

Procedure.

1. Make a model of methane, CH_4.

2. Convert it to a model of methyl chloride, CH_3Cl. Does it matter which of methane's four hydrogens is replaced by a chlorine? If not, the hydrogens are equivalent. (That is, they have identical environments with respect to the other atoms adjacent to themselves.) Are the three hydrogens in methyl chloride equivalent?

3. Convert your model of methyl chloride into dichloromethane, CH_2Cl_2. Are the two hydrogens in this model equivalent?

4. Make a model of ethane, CH_3CH_3. Are the six hydrogens equivalent?

5. Make a model of ethyl chloride, CH_3CH_2Cl. Are the remaining five hydrogens equivalent?

6. Make models of the possible dichloroethanes, $C_2H_4Cl_2$. Write their condensed structural formulas on the Report Sheet. These different arrangements of the same empirical formula are called isomers.

7. Make a model of propane, C_3H_8. Are isomers possible? If so, draw their condensed formulas.

8. Replace one of the hydrogens of the propane model with a chlorine atom. Make models for and, on the Report Sheet, write the condensed formulas for all of the possible isomers of the formula C_3H_7Cl.

9. Pooling two sets of model parts with a neighbor make models of the possible isomers of C_4H_{10}. Write both full and condensed structural formulas.

Examine each structure to discover sets of equivalent hydrogens. (Use the models you made.) Using circles or squares or triangles as needed, identify the hydrogens belonging to each set. Draw these figures around individual hydrogens above in the full structures you wrote.

10. Now replace one of the hydrogens on the butane model with a chlorine. Determine how many isomers of C_4H_9Cl exist and write their condensed structures on the Report Sheet.

11. What relation exists between the number of sets of equivalent hydrogens in an alkane (such as C_3H_8 or C_4H_{10}) and the number of isomers having one atom substituted for a hydrogen (a monosubstituted derivative) of that alkane?

EXPERIMENT 27 STUDIES IN STRUCTURAL ORGANIC CHEMISTRY: ISOMERISM

1. Are the three hydrogens in methyl chloride equivalent? _____

2. Are the two hydrogens in dichloromethane, CH_2Cl_2, equivalent? _____

3. Are the six hydrogens in ethane, CH_3CH_3, equivalent?_____

4. Are the remaining five hydrogens of ethyl chloride, CH_3CH_2Cl, equivalent?

_____

5. Dichloroethanes, $C_2H_4Cl_2$: write their condensed structural formulas below.

6. Propane, C_3H_8: are isomers possible? If so, draw their condensed formulas.

7. Write the condensed formulas for all of the possible isomers of the formula C_3H_7Cl.

8. Write both full and condensed structural formulas of all isomers of C_4H_{10}.

Using circles or squares or triangles as needed, identify the hydrogens belonging to each *set* of equivalent hydrogens.

9. Determine how many isomers of C_4H_9Cl exist and write their condensed structures below.

10. What relation exists between the number of *sets* of equivalent hydrogens in an alkane (such as C_3H_8 or C_4H_{10}) and the number of *isomers* having one atom substituted for a hydrogen (a monosubstituted derivative) of that alkane?

CHAPTER 10

Hydrocarbons

Hydrocarbons are compounds composed of only carbon and hydrogen. They have distinctive properties, which we will investigate. Before we begin, however, it is useful for us to make models of compounds in order to build the mental connections that will allow us to look at a formula, a listing of C's and H's if you will, and understand its three-dimensional implications.

EXPERIMENT 28 HYDROCARBONS

WARNING: In experiments with organic substances assume that they are all extremely flammable. Be very careful about the use of a laboratory burner. Read the following section before coming to the laboratory.

Other chemicals used in this experiment are hazardous in other ways. Sulfuric acid is one.

WARNING: Sulfuric acid attacks skin and fabrics swiftly. If you spill it on yourself (or others!), don't wait. Rinse it with cold water at once—with copious amounts. Daub the spot with solid, powdered sodium carbonate or bicarbonate. If a blister forms, go to the health service. Let your instructor know you are going.

WARNING: Never pour concentrated sulfuric acid directly into a sink or disposal trough. Some may spatter out when a faucet is turned on. Pour waste sulfuric acid into a large amount of cold water to dilute it. Then pour it into the drain hole—directly into the drain hole, not elsewhere. Then let the faucet run a while.

Bromine is another dangerous chemical. As a pure, liquid element it can cause severe, painful burns. Its vapors are very irritating to the nose and throat. Use even its solutions only at the hood.

To make bromine safer, it is diluted with methylene chloride, a solvent inert to bromine and which functions only as that—a solvent, not a reactant in the tests you will do. Keep the tube away from your other hand.

Warning: Dispose of all organic wastes as directed by the instructor.

A. Concentrated Sulfuric Acid Test

What you can expect to see:

1. With alkanes, cycloalkanes, toluene, alkyl halides, and aryl halides, the sulfuric acid will form a separate layer. No heat is evolved; no discoloration is

observed; no apparent reaction takes place. (Should these be contaminated by substances in families reactive toward sulfuric acid, then you will see something else, of course. See category 2 next. The extent of change depends on the amount of contamination.)

2. With alkenes, alkynes, and any organic compound containing oxygen, nitrogen, or sulfur, heat will evolve. The mixture may turn black. Blackish solids may separate. No separate layer may be detected

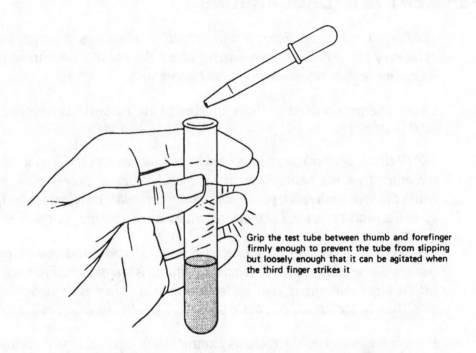

Grip the test tube between thumb and forefinger firmly enough to prevent the tube from slipping but loosely enough that it can be agitated when the third finger strikes it

Fig. 27 Mixing the contents of an open test tube while leaving one hand free to add the reagent.

B. Bromine Test

What you can expect to see:

1. With alkanes, alkyl benzenes and most organic compounds whose molecules do not have carbon-carbon double or triple bonds all you will see at first is the dilution of the bromine color. (However, see category 3, below.)

2. With most alkenes and alkynes, the deep brown color of the reagent disappears almost instantaneously. It's very striking, and quite different from merely diluting the color.

3. You may eventually see a fuming in the test tube, particularly if the room air

is humid or if you blow moist breath across the tube. The fuming is caused by hydrogen bromide gas interacting with moisture. Many compounds in category 1, above, will display this behavior a minute or two after the initial bromine color has been merely diluted. As fuming occurs, the diluted color disappears, too. As we discuss next, hydrogen bromide arises from a substitution reaction. You can make it commence by holding the test solution up to sunlight, which initiates the reaction. You can detect the evolution of HBr by holding a strip of moist blue litmus paper in the neck of the test tube (but not touching the sides of the test tube). The litmus turns red from the liberated hydronium ion.

$$HBr + H_2O \rightleftharpoons H_3O^+ + Br^-$$

The Chemistry of the Tests. Both sulfuric acid and bromine add to carbon-carbon double bonds:

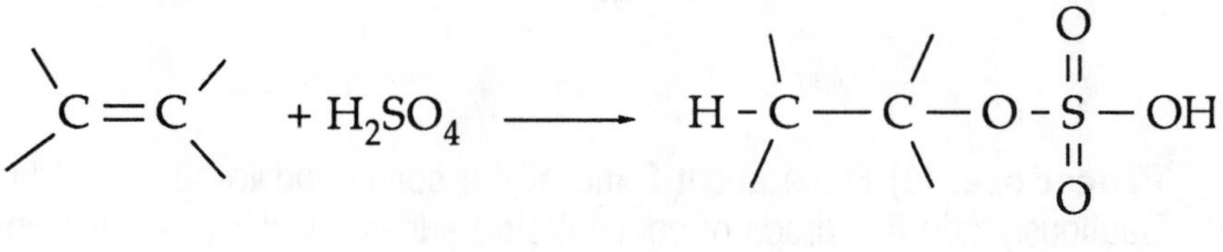

Alkyl hydrogen sulfate

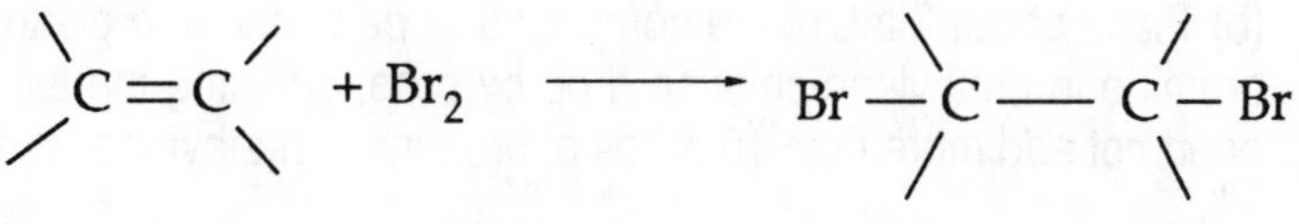

Dibromoalkane

Alkyl hydrogen sulfates are strong acids and are polar enough to dissolve in sulfuric acid. Sulfuric acid catalyzes other reactions, some of them complex, which produce colored products.

Dibromoalkanes are colorless (or pale yellow) compounds. Bromine attacks alkanes slowly, by a substitution process that results in an acid by-product:

$$R{-}H \;+\; Br_2 \longrightarrow R{-\!\!-}Br \;+\; H{-}Br \uparrow$$

| Alkane | | Alkyl bromide (colorless) | Hydrogen bromide |

Alkane   Alkyl bromide   Hydrogen
                                (colorless)   bromide

Organic compounds containing oxygen, nitrogen, or sulfur will be bases toward concentrated sulfuric acid, because these atoms will nearly always have unshared, outer-shell electrons, whatever their mode of inclusion in the organic substance. The acid will donate one of its two protons to the base, and (at least initially) ionic species that are soluble in the sulfuric acid will be produced. Subsequent reactions will nearly always occur, and colored products often form as heat is evolved. To illustrate, using a general alcohol:

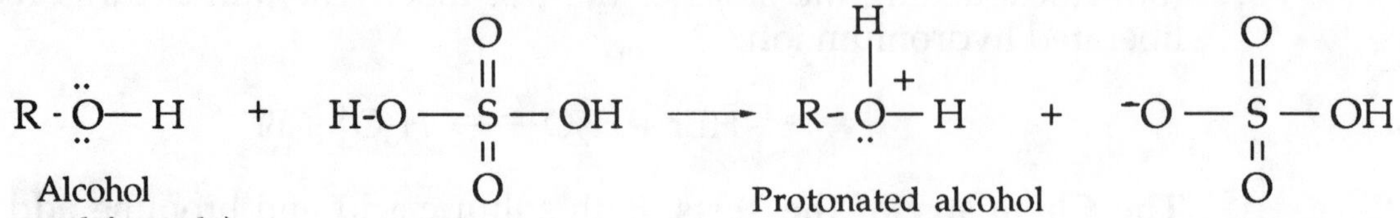

Alcohol (general formula) stronger base; Sulfuric acid stronger acid; Protonated alcohol weaker acid; Hydrogen sulfate ion weaker base

Procedure. (a) Place about 2 mL of the compound to be tested in a clean, dry test tube. Cautiously add five drops of concentrated sulfuric acid. Agitate the contents of the tube after each drop. (Figure 27 shows how to do this.) Do the test on the "knowns" listed on the Report Sheet.

(b) Place about 2 mL of the compound to be tested in a clean, dry test tube. Add a solution of bromine in methylene chloride drop by drop, agitating the tube as you do so (Figure 27). You need not add more than 10 drops of bromine in methylene chloride.

(c) The unknowns are in dropper bottles labeled A, B, and C. Test each with sulfuric acid, and with bromine, like you did for the "knowns." Record your results. Based on what you see happen, determine the chemical family of each.

EXPERIMENT 28 HYDROCARBONS

Knowns	Describe what you saw	
	A. Conc. H_2SO_4	B. Br_2 test
Alkane:_____ Condensed structure:		
Alkene:_____ Condensed structure:		
Aromatic:_____ Condensed structure:		
Alcohol:_____ Condensed structure:		
Unknowns: Which family is represented by each bottle? A._____ B. _____ C. _____	A. B. C.	A. B. C.

There was an emphasis in the directions for the concentrated sulfuric acid test on using dry test tubes. Explain how the presence of a drop or two of water would confuse the results.

EXPERIMENT 29 EXERCISES IN STRUCTURAL ORGANIC CHEMISTRY: HYDROCARBONS

Distributed about the laboratory are several "ball-and-stick" models of simple organic compounds. Working in assigned teams and in a designated hraffic pattern, spend a few minutes studying each model, writing its shructural formula, and answering the question or questions on the card by each model. (You may refer to your textbook if your instructor agrees.)

Write orderly structural formulas as they are conventionally written or typed. Do not try to duplicate the particular conformation or twist in space that the model may have on the desk. Do not handle the models in such a way that they come apart. Should that happen, inform your Instructor, unless the repair can be made simply and correctly.

Structure and Name	Answers to Questions
1.	
2.	
3.	
4.	

5.

6.

7.

8.

9.

10.

CHAPTER 11   Alcohols, Ethers, Thioalcohols, and Amines

EXPERIMENT 30 STRUCTURE AND SOLUBILITY (POLARITY)

WARNING: Most chemicals used in this experiment are very flammable. There must be no flames anywhere in the laboratory. If possible, work at the hood. Otherwise, have good ventilation.

Prelab Exercise. Using the type of structural formulas specified by your instructor of isopropyl alcohol, propylene glycol, and glycerol, draw diagrams showing how they can hydrogen bond to themselves and how the increased possibilities for hydrogen bonding of propylene glycol and of glycerol could cause a network of bonding that holds many molecules together. When you get to the lab, compare the relative viscosities of these three substances and note how they do tend more and more to "stick" to adjacent molecules as the number of—OH groups on a single molecule increases. Then consider that this network of bonds must have work expended on it to break the bonds before the molecules can hydrogen bond to water and thus dissolve.

Prelab Exercise.

Procedure. Test the solubilities of the compounds listed on the Report Sheet in two solvents: water and diethyl ether. When using ether, the test tube must be dry. Add 10 drops of each of the compounds listed on the Report Sheet to 1 mL of solvent. Mix the contents of the tube (as in Fig. 27). If the substances do not seem to be dissolving with the method of Fig. 27, use a stirring rod before you declare them immiscible with each other. Consider how much "work" you must expend stirring to get sugar to break hydrogen bonds between its molecules so it can hydrogen bond to water (and dissolve in coffee or tea.)

Look carefully for two layers. They can be difficult to see when the liquids are colorless. When a test is completed, discard the waste liquids at the hood as directed by the instructor. The ether solutions should not be discarded into drains at the laboratory benches.

EXPERIMENT 30 STRUCTURE AND SOLUBILITY (POLARITY)

Compound Tested			Solubility in:	
Name	Structure	Family	Water	Ether
Pentane				
Diethyl ether				
Isopropyl alcohol				
Propylene glycol				
Glycerol				

Explanations

1. Explain why pentane is less soluble in water than isopropyl alcohol. Use drawings.

(Questions continue on the back of the page.)

2. Explain how glycerol behaves as it does in diethyl ether.

3. Which of the compounds you tested was the MOST polar? Draw its structure showing where the polarity originates in the molecule.

EXPERIMENT 31 OXIDATION OF ALCOHOLS (PRIMARY, SECONDARY, TERTIARY)

In this experiment you will study a method whereby aldehydes or ketones may be synthesized from appropriate alcohols. The experiment, therefore, provides a study of the behavior of 1°, 2°, and 3° alcohols toward a strong oxidizing agent (sodium dichromate). Methods for isolating the products of the oxidations in pure form are too involved for our study. That a reaction has occurred, however, can readily be observed. First, the oxidizing agent, the dichromate ion, $Cr_2O_7^{2-}$, is bright orange. When it reacts as an oxidizing agent in this experiment, it is reduced to the brilliant green chromium(III) ion, Cr^{3+}. Hence, if the alcohol studied is oxidized by the reagent, you will observe a dramatic change in the color of the solution.

Second, the aldehyde and ketone you will prepare have odors sharply different from their parent alcohols. Therefore, if oxidation occurs, you will notice a change in odor. (Use caution whenever you check the odor of any compound. Do not breathe deeply. The odor of a bottled sample may usually be noted simply by smelling the stopper or bottle cap. When checking the odor of a solution, slowly and cautiously bring your nose to the edge of the container (Fig. 28).

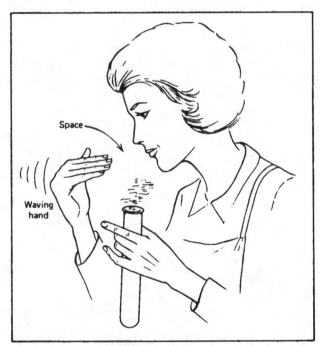

Fig. 28 Be careful in checking odors. Waft vapors to the nose with your hand.

Procedure. Study the oxidation of each alcohol in the following way. Perform the experiment with ethyl alcohol, isopropyl alcohol, and t-butyl alcohol.

Place 3 mL of 5% sodium dichromate in a small beaker or Erlenmeyer flask. Cautiously add 1 mL of concentrated sulfuric acid. If a precipitate appears, swirl the mixture until it dissolves. Cool the solution. (Hot sulfuric acid easily dehydrates alcohols to alkenes, which will be oxidized.)

Slowly add 2 mL of the alcohol to be tested. If the color of the solution changes to green, note its odor and compare that odor with the odor of an authentic sample of the aldehyde or ketone you would expect to form under the conditions.

Since aldehydes are very easily oxidized, you will get a mixture of odors from the oxidation of ethyl alcohol. Put the name and formula of a second possible product of oxidation on the Report Sheet.

EXPERIMENT 31 OXIDATION OF ALCOHOLS

Alcohol used	Structure	(1°, 2°, or 3°)	Structure of expected oxidation product
Ethyl alcohol			
Isopropyl alcohol			
t-Butyl alcohol			

Describe what you observed (color change and odor).

Ethyl alcohol:

Isopropyl alcohol:

t-Butyl alcohol:

If, with any of the three alcohols, no reaction was observed, explain why.

EXPERIMENT 32 PROPERTIES OF PHENOLS

Aromatic alcohols, or phenols, have special chemical and physical properties not shown by the aliphatic alcohols we have studied earlier. The fact that phenol (hydroxybenzene) is also called carbolic acid is an indication of the behavior of this compound with a base.

32 A. Solubility, and Acidic Nature of Phenol.

Procedure. Using a spatula or forceps, select two small crystals of phenol as nearly the same size as possible; place them in two test tubes. Cautiously note the odor. Add 5 mL of distilled water to one and 5 mL of dilute (3 M) sodium hydroxide to the other. Swirl the two tubes equally and note the relative speed with which the crystals dissolve. Record your observations. Compare the solubility of phenol in water with the solubility of a hydrocarbon e.g., pentane in water.

32 B. Tests for Phenols.

1. Bromine water test.

Procedure. Take 1 mL of the water solution of phenol and dilute it with 5 mL of distilled water. To this dilute solution, add 1 mL of bromine water. The precipitate that forms is tribromophenol. To prove that this is a substitution product and that HBr is being evolved, hold a piece of moist blue litmus in the mouth of the test tube (but not touching the walls). Does it turn red? The substitution of bromine on the benzene ring of benzene itself (Experiment 28) is achieved only in a nonpolar solvent and in the presence of bright light or an iron catalyst. The presence of the - OH group on the benzene ring as in phenol thus dramatically changes the reactivity of the ring hydrogens, as well as the solubility of the compound in water.

2. Ferric chloride test

Procedure. Take 1 mL of the aqueous phenol solution and dilute it with 5 mL of distilled water. To this add one drop of ferric chloride solution. The production of a purple color with this reagent is often used as a test for the phenol group. To supplement this observation, add a drop of ferric chloride solution to 2 or 3 mL of a salicylic acid solution.

32 C. Acidity of Phenol

Procedure. To the remainder of the aqueous solution of phenol, add two or three drops of universal indicator solution. If universal indicator is not available, a crystal of phenol can be transferred by a spatula or forceps directly onto moist blue litmus paper to show acidity—not strength.

EXPERIMENT 32 PROPERTIES OF PHENOLS

32 A. Describe the odor of phenol (is it familiar?)

How did the solubility of phenol in water compare with the solubility of the hydrocarbon pentane in water in Experiment 30? Explain your answer, preferably with a hydrogen bonding diagram.

32 B. Tests for Phenols

1. Assuming that the bromine substitutes on the two carbons ortho to (adjacent to, on each side of) the hydroxyl group, and on the carbon directly opposite (para to) the hydroxyl, write the equation, showing the structures of the reactants and the products, and naming them.

2. Does salicylic acid also give the ferric chloride test? _____ Find the structure of salicylic acid in Experiment 39, write it here, and circle the reactive group.

32 C. Is the aqueous phenol solution acidic, neutral, or basic?

Review the meaning of strong and weak acids. Is phenol a strong acid (e.g., like hydrochloric acid) or a weak acid?

How did you decide?

Write the reaction, using structural formulas, between phenol and NaOH, showing the products as ions (plus water). Name the structures.

EXPERIMENT 33 BASICITY OF AMINES: A COMPARISON WITH AMMONIA

In many of their properties, organic amines resemble ammonia. In this experiment, you will study and compare the basicity of an amine, triethylamine, with that of ammonia. You will also review how molecules whose structures contain no obvious potential hydroxide ions (or hydrogen ions) can still upset the balance of the trace amounts of these ions in neutral water and make the solution basic (or acidic). You will also see how the solubility of an organic amine can be altered quickly and easily at room temperature merely by changing the pH of the medium (by adding acid or base).

$$CH_3CH_2 \overset{\displaystyle \overset{..}{N}}{\underset{\displaystyle CH_2CH_3}{\big|}} {}^{{}_{\cdots\cdots}} CH_2CH_3 \qquad\qquad H \overset{\displaystyle \overset{..}{N}}{\underset{\displaystyle H}{\big|}} {}^{{}_{\cdots\cdots}} H$$

Triethylamine   Ammonia

You will be more popular with your labmates if you work at the hood with these chemicals.

Procedure.

1. Place 1 mL of 5 M aqueous ammonia in one test tube and about 10 drops of triethylamine in another. Describe their odors (care!), emphasizing likenesses or differences. Place all answers to these questions on the Report Sheet.

2. With a glass rod (rinsed well between uses), touch a drop of each substance to a piece of moist red litmus paper and record your observations.

3. Add about 1 mL of water to the triethylamine and note its solubility. Test the water solubility of the ammonia also.

4. To both the water solutions of ammonia and triethylamine, add 3M HCl (hydrochloric acid), drop-by-drop with constant agitation of the tubes, until the odors have disappeared (or at least, virtually so.) Be sure any material that might be clinging to the walls of the tubes or that is simply in the vapor phase above the surface of the liquid is given a chance to interact with the added acid. What reactions occurred that converted ammonia and triethylamine into essentially odorless substances and, in the case of triethylamine, into a water-soluble substance? (If odors persist it may be necessary not only to add additional acid but also to stopper the tubes with tight-fitting corks and shake them vigorously but <u>carefully!</u>)

5. Next, add to each tube about as much 3 M sodium hydroxide as you used of 3 M hydrochloric acid. Describe any changes in odor.

6. What else did you observe in the tube originally containing triethylamine after you added sodium hydroxide?

Name ______________________________ Partner ______________________________

Section __________ Date ___________ Due Date ____________ Score ______________

EXPERIMENT 33 BASICITY OF AMINES: A COMPARISON WITH AMMONIA

1. (Odors and other observations):

2. Litmus Test: Ammonia________________________________

Triethylamine____

The effect on moist litmus by the triethylamine is due to hydrolysis. What ion taken from a water molecule did molecules of triethylamine *bind*? _______ What ion was left in excess therefore, to account for the color change in the litmus? _________ How does this behavior resemble that of ammonia molecules in water? *Write a chemical equation for each as part of your answer.*

3. Solubility Test: Ammonia________________________________

Triethylamine ____

4. Write an equation showing what happened, first to the ammonia, then to the triethylamine, when acid was added. Label all compounds.

5. (Odors and other observations):

Write ionic equations, using structural formulas, to show the formation of substances that would account for the properties observed when the sodium hydroxide was added. Place the names of the products beneath these structures.

EXPERIMENT 34 EXERCISES IN STRUCTURAL ORGANIC CHEMISTRY: ALCOHOLS, ETHERS, THIOALCOHOLS, AND AMINES

Distributed about the laboratory are a few ball-and-stick models of molecules belonging to the families of alcohols, thioalcohols, amines, and ethers. Examine them; write neat, condensed structural formulas in the spaces provided below each number and the name of the compound. Answer the number-coded question by each model.

Structure and Name	Answers to Questions
1.	
2.	
3.	
4.	

5.	
6.	
7.	
8.	
9.	
10.	

CHAPTER 12   Aldehydes and Ketones

EXPERIMENT 35 OXIDATION OF CARBONYL COMPOUNDS

You will now study the behavior of representative carbonyl compounds from three families—aldehydes, ketones, and carboxylic acids—toward oxidizing agents of various strengths. Your instructor will specify which compounds to use in each case.

35 A. Action of Air as an Oxidizing Agent

Aldehydes are sometimes so easy to oxidize that exposing them to air at room temperature will carry them to the carboxylic acid stage. Benzaldehyde is a very satisfactory material to use for this demonstration, since it is a volatile oil with a pleasant odor, which will be converted during the course of the laboratory period to benzoic acid, a white solid with no odor.

Procedure. Put a drop of benzaldehyde on a watch glass and spread it around on the glass in a thin film. Note and record the odor, remembering it must be volatile for this odor to reach your nose. Fortunately, it is not so volatile that it will evaporate before most of it is converted to benzoic acid. Watch the white solid appear mixed in with the oil, and by the end of the period the oil should be completely converted to colorless crystals, with no odor. While this change is occurring, you can be working out the rest of the exercise.

35 B. Action of a Mild Oxidizing Agent

Tollens' Reagent contains silver ions, and a positive Tollen's test depends on the presence of an aldehyde group. As the aldehyde is oxidized to a carboxylic acid, the silver ions are reduced to metallic silver, which forms a mirror-like deposit on the walls of the test tube.

Procedure. Clean three test tubes thoroughly with detergent solution and a brush. Rinse them well with tap water, with a final rinse with distilled water. Place 2 mL of 5% silver nitrate solution into each tube. To each tube, add 1 drop of 3M NaOH. Now to each tube add 2% ammonium hydroxide drop-by-drop until the brownish precipitate of silver oxide (Ag_2O) just dissolves as it forms soluble $Ag(NH_3)_2^{+}$ ions (with OH$^-$ present). Be patient, the precipitate will dissolve; be careful not to add an excess of ammonium hydroxide, however, since this would decrease the sensitivity of the test. To the first tube, add 2 drops of 10% glucose, to the second tube add 2 drops of acetone, and to the third tube add 2 drops of formaldehyde or an equivalent amount of another aldehyde.

Mix the contents of each tube, and place them in a hot water bath (about 60 °C). Observe, and record any changes. Discard the waste solutions in the sink, using plenty of running water to rinse them down the drain. Scrub the test tubes again with detergent and rinse. If any silver deposit remains, take the tube to the hood and use a few drops of concentrated nitric acid to dissolve it. Rinse this well with water also. If the residue dries, it becomes explosive.

35 C. Action of Strong Oxidizing Agent

You used dichromate ion in Experiment 31. Here you will use another common strong oxidizing agent, permanganate ion, MnO_4^-. (It can be used in the form of an aqueous solution of its potassium salt, $KMnO_4$, called potassium permanganate.) The permanganate ion gives a deep reddish-purple color to water, but when it reacts as an oxidizing agent, it is reduced to a brown, sludgelike precipitate, MnO_2, called manganese dioxide. That change, then, is what you see when permanganate ion reacts as an oxidizing agent—a reddish-purple solution becomes lighter in color, or colorless, and some brown sludge separates.

Procedure. Obtain 15 mL of 1% potassium permanganate. Distribute it among three clean test tubes, about 5 mL in each. To one tube, add 5 drops of an aldehyde. To the second tube, add 5 drops of a ketone; and to the third, add 5 drops of a carboxylic acid. Record your observations on the Report Sheet, and answer the questions posed there.

Name ______________________________ Partner ______________________

Section __________ Date ___________ Due Date _____________ Score _______________

EXPERIMENT 35 OXIDATION OF SOME CARBONYL COMPOUNDS

35 A. Write the equation for the air oxidation of benzaldehyde to benzoic acid showing structural formulas of organic compounds.

35 B. Tollens' Test Observations:

glucose____

acetone ____

formaldehyde____

The presence of what functional group in the glucose molecule is indicated by this test? Write the structure of that group below.

(Continued on the back of the page.)

35 C. Strong Oxidizing Agent

Family	Name of compound	Condensed structure	Observed behavior toward permanganate	Oxidation product
Aldehyde				
Ketone				
Carboxylic acid				

For the reactions above, where you observed a chemical change, write the name and the structure of the organic product(s). (Write the name beneath the structure(s))

CHAPTER 13

Carboxylic Acids and Their Derivatives

EXPERIMENT 36 CARBOXYLIC ACIDS AND THEIR SALTS

The presence of the carbonyl on the carboxylic acid molecule affects the behavior of the adjacent hydroxyl group. The hydrogen is no longer tightly bonded to the oxygen as it was in the corresponding alcohol. Instead, the oxygen-hydrogen bond is weakened, and the hydrogen can be removed by a substance that can accept protons. Thus these compounds are Bronsted acids:

$$R-\overset{\overset{\displaystyle O}{\|}}{C}-OH + OH^- \rightleftharpoons R-\overset{\overset{\displaystyle O}{\|}}{C}-O^- + H_2O$$

      Acid   Base

In this experiment, we will investigate the properties of these compounds.

Procedure.

1. Put a small amount of salicylic acid in a test tube and add 5 mL of water. Stir well. Does it dissolve in the cold water?

2. Carefully heat the mixture until it starts to boil. Does it dissolve in hot water?

3. Cool the solution to room temperature by holding the lower end of the tube in cold water. What happens? (If no observable change occurs, insert a glass rod in the solution and rub the inner wall of the tube.)

4. Next add 3 M sodium hydroxide to the mixture drop-by-drop, agitating the tube until the crystals dissolve. Write the equation for the reaction of salicylic acid with sodium hydroxide on the Report Sheet, showing structures of the organic substances. Under each formula, write its name, and whether or not it is soluble in water. If any of these are ions, you must include the charge.

5. Add to the solution as much 3 M hydrochloric acid as you used 3 M sodium hydroxide; then add several more drops of the acid. What happened? (What did the hydrochloric acid do to the salicylate ion?)

EXPERIMENT 36 CARBOXYLIC ACIDS AND THEIR SALTS

1. Observation:_____

2. Observation:_____

3. Observation:_____

4. Equation for what was done in part 4:

5. Observation:_____

6. Write the net ionic equation for the reaction of the salicylate ion (in part 5) with HCl, again showing structures, names, and charges, and indicating the water solubility for each substance.

7. Why is the salicylate ion more soluble in water than salicylic acid?

8. (Optional) If an organic base had been used instead of an organic acid in this experiment, for example, aniline, what reagent could be added to increase its solubility in water?

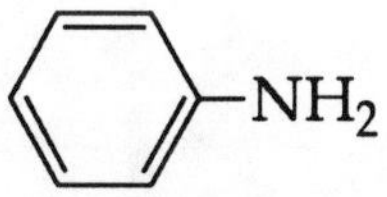

$-NH_2$ (Aniline)

Write the structure of the new water-soluble substance:

EXPERIMENT 37 ESTERIFICATION: SYNTHESIS OF METHYL SALICYLATE (OIL OF WINTERGREEN) AND OTHER PLEASANT ODORS

In this experiment you will demonstrate how an ester can be made by the interaction of a carboxylic acid and an alcohol, in the presence of an acid catalyst, sulfuric acid. You will use one or more of the combinations listed on the Report Sheet, as designated by your instructor. If time permits, your instructor will show you how the sulfuric acid acts as a catalyst.

$$R-\overset{\overset{\displaystyle O}{\|}}{C}-OH \ + \ R'-CH_2-OH \longrightarrow R-\overset{\overset{\displaystyle O}{\|}}{C}-O\text{-}CH_2\text{-}R' + H_2O$$

Evidence of reaction in these esterification reactions is twofold. One clue is a change in solubility. The alcohol in each case is water soluble, and in the quantities used here, so are the acids (salicylic is marginally so). The products are an ester and water, but since the ester is not water soluble, you can expect to see two layers. The second clue is a distinctive change in odor. Note very cautiously the odors of the alcohol and the acid.* They are usually sharp and extremely unpleasant. The products, however, have pleasant odors, some reminiscent of certain fruits. If the reaction is too incomplete, however, the odors of the unreacted alcohol and/or acid may confuse the olfactory picture. *OPTIONAL!

Procedure. Prepare a hot water bath, and maintain the temperature at 60-70 °C.

Use dry test tubes. Place 3 mL of the assigned alcohol in the test tube. Dissolve in it (with stirring) about 0.5 g of a solid acid, (e.g., salicylic acid) or about 1 mL of a liquid acid. When solution is complete, slowly add 10 drops of concentrated sulfuric acid (care!) drop-by-drop, with stirring.

Place the test tube in the water bath, and allow it to remain there for at least 10 minutes. Some of the esters form quickly, while others require more time. If, at the end of the 10 minutes, the odor of either the acid or the alcohol is still obvious, increase the temperature another 10 degrees, and heat for another 5 minutes. When you think that there has been a sufficient reaction, pour the contents of the tube into 20 mL of hot water contained on a small beaker. Again, cautiously note the odor, and describe it on your Report Sheet.

EXPERIMENT 37. ESTERIFICATION: SYNTHESIS OF METHYL SALICYLATE (OIL OF WINTERGREEN) AND OTHER PLEASANT ODORS

In this experiment you will demonstrate how an ester can be made by the interaction of a carboxylic acid and an alcohol in the presence of small amount of sulfuric acid. You will use one member of the homologous series of the aliphatic acids designated by your instructor. This reaction, your instructor will show you how the esters are made and identified.

EXPERIMENT 37 ESTERIFICATION

Complete the following table of esterification reaction equations before coming to the laboratory. You will be using one or more of these combinations, as your instructor directs. After reactions are complete, fill in the descriptions of the odors from your ester(s), and from those around you who tried other combinations.

EQUATION	ODOR OF ESTER
METHYL ALCOHOL + SALICYLIC ACID	
ETHYL ALCOHOL + BUTYRIC ACID	
n-PENTYL ALCOHOL + ACETIC ACID	
ISOPENTYL ALCOHOL + ACETIC ACID	
n-PENTYL ALCOHOL + BUTYRIC ACID	
n-PENTYL ALCOHOL + PROPIONIC ACID	
(Continued on back of page.)	

OTHER:	
OTHER:	

EXPERIMENT 38 ASPIRIN (Acetylsalicylic Acid, "ASA")

Acetylsalicylic acid, commonly called aspirin, is an ester of acetic acid and salicylic acid (the latter acting as the "alcohol"). Although esters of acetic acid may be made by direct interaction of acetic acid with an alcohol or phenol, chemists commonly employ a relative of acetic acid, acetic anhydride, as a substitute acetylating agent. It forms acetate esters far more quickly than does acetic acid directly. Normally, sulfuric acid is a catalyst. Acetic anhydride acetylates a phenol as follows:

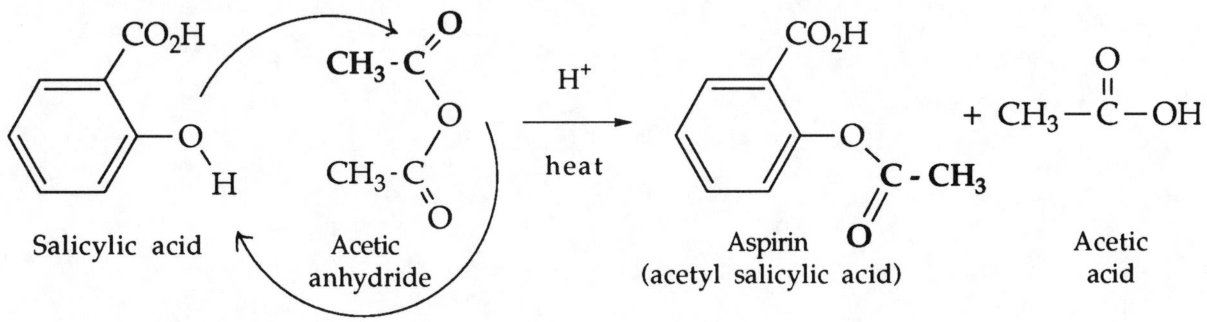

Salicylic acid   Acetic anhydride   Aspirin (acetyl salicylic acid)   Acetic acid

Equipment.

a 250 mL or 400 mL beaker for a water bath (start this heating before doing anything else!),

3-4 clean, dry test tubes,

a 60° short-stem funnel,

stirring rod,

filter paper,

*medicine droppers,

*a spot plate.

*(For next laboratory period.)

Procedure.

1. Write your initials in pencil on the filter paper and weigh it to the nearest 0.001 g; record this on your report sheet. Obtain 1 gram of salicylic acid, weighing it and recording its mass to the nearest 0.001 g on your report sheet.

2. Mix together in a dry test tube 2 mL of acetic anhydride and 1 g of salicylic acid. Add 2 drops of concentrated sulfuric acid.

3. Now place the test tube in the water bath and stir the mixture vigorously while you heat it, until the solid has dissolved.

4. Set the tube aside to cool. If no crystals appear when the tube is cooled to room temperature, scratch the inner wall of the tube with the glass rod. When crystallization is complete, add 10 mL of cold water, stir, and collect the solid on the filter paper in the funnel. Rinse the solid on the filter paper with 2 or 3 small (5 mL) portions of cold water, letting the water filter through each time.

5. Gently lift the filter paper with the product out of the funnel and spread it out on a piece of paper toweling. Set it in the place designated by your instructor to dry until the next laboratory period. The product is acetylsalicylic acid (aspirin).

Name ______________________________ Partner ______________________________

Section __________ Date __________ Due Date ____________ Score ______________

EXPERIMENT 38 ASPIRIN

Data and Calculations: (Items marked with an asterisk * are done in the second laboratory period, after the product is thoroughly dry.)

Mass of filter paper ______________________________

*Mass of filter paper plus product ______________________________

*Mass of acetylsalicylic acid (aspirin) ______________________________

Mass of salicylic acid used ______________

Moles of salicylic acid used ______________

Moles of acetylsalicylic acid expected (theoretical yield) (Refer to the equation.)

Mass of ASA expected (theoretical):

____

*% yield = (mass of ASA(actual) x 100) ÷ mass of ASA(theoretical)

________________________ %

*Test for purity: (Ferric chloride test)

(a) Dissolve a small amount of your product in 5 mL of water. Add 1—2 drops of ferric chloride solution. The reagent detects phenol groups; a positive test is a violet color. Observations:

(b) Repeat the test using a small piece of a commercial aspirin tablet. Observations:

Which of the two products seem to contain more unreacted phenol?

What is the significance of these unreacted phenol groups in a sample of aspirin, whether in yours or the commercial sample?

Now place a fragment of a commercial aspirin tablet on a spot plate. Add a drop or two of water, and a drop of iodine reagent. Was starch present?

What do you think is the purpose of starch in the tablet?

EXPERIMENT 39 SAPONIFICATION: THE SAPONIFICATION OF METHYL BENZOATE

39 A. Pre-experimental Work. Complete parts 1-7 before coming to the laboratory.

1. Write the equation for the saponification of methyl benzoate showing all structures. Beneath each structure write its name, and in parentheses beneath each name write "sol." if the substance is soluble in water and "insol." if it is not.

Methyl benzoate

Methyl benzoate is used in this experiment so you can the course of the reaction. Consider the properties following compounds involved in the reaction:

Methyl benzoate:	a water-insoluble oil
Sodium benzoate:	a water-soluble salt-like solid
Methyl alcohol:	a water-soluble liquid alcohol
Benzoic acid:	a white solid, appreciably soluble in hot water, but insoluble in cold water.

On the basis of these properties, answer the following questions about expected observations before you carry out the experiment.

2. Define the term "homogeneous":

3. When methyl benzoate is added to aqueous sodium hydroxide, will a homogeneous solution form immediately (assume that saponification is a slow reaction)? Why?

4. When saponification is complete, will the solution appear homogeneous? Why?

5. Would the rate of saponification be hastened by frequent shaking of the reacting mixture? Why?

6. Consider now the nature of the initial products of the saponification. What chemical change will occur if the solution resulting from the reaction were acidified with 10% hydrochloric acid? Write an ionic equation.

7. If the acidification is done to the hot solution resulting from the reaction, will you necessarily see evidence for the chemical change in question 6? Why?

8. What might you see if you cooled the acidified solution in question 7?

39 B. Experimental Procedure.

Procedure. Place 3 drops of methyl benzoate in a clean test tube and add 2 mL of water and 12 drops of 10% sodium hydroxide. Mix the contents of the tube and place the tube in a beaker of boiling water. Heat the tube in the boiling water bath for half an hour or longer. At frequent intervals stir the contents of the tube vigorously. When you judge the reaction to be complete (see question 3), remove the tube from the water bath and add 12 drops of 3 M hydrochloric acid. Mix the contents of the tube well and test to see that the solution is acidic; if not, add more acid, by drops, until it is.

9. What visual change, if any, do you observe?

10. If no change was observed, allow the tube to cool to room temperature and record any visual changes that occur. From questions 1-8, what expected observations, if any, did not materialize?

EXPERIMENT 40 ALKALINE HYDROLYSIS OF AN AMIDE: ACTION OF AQUEOUS SODIUM HYDROXIDE ON BENZAMIDE

40 A. Pre-experimental Work. Complete parts 1-4 before coming to the laboratory.

In this experiment you will see how an amide can be hydrolyzed by the action of hydroxide ion. Benzamide is used because it makes it possible for you to watch the course of the reaction.

1. *Write an equation* that symbolizes the expected behavior of benzamide toward aqueous sodium hydroxide. Below each condensed structure or formula write the name of the compound and state if it is water-soluble. (Benzamide is not soluble in water.)

$$
\underset{\substack{\text{Benzamide} \\ \text{(Insoluble)}}}{}\quad
\begin{array}{c} O \\ \| \\ -C - NH_2 \end{array} \;+\; \underline{} \;\longrightarrow\; \underline{} \;+\; \underline{}
$$

Benzamide
(Insoluble)

2. As the reaction proceeds, will you expect to see the development of a homogeneous solution? Why?

3. Will you expect to note any change in the odor of the contents of the tube (be most cautious in testing) as the reaction proceeds? What substance(s) would you be able to detect this way?

4. Will frequent agitation of the contents of the tube help to hasten the reaction? Why?

40 B. Procedure. Place approximately 0.5 g of benzamide in a clean test tube, add 3 mL of 3 M sodium hydroxide, and place the tube in a beaker of boiling water. Stir the mixture frequently. After a few minutes the solid will have melted and it will appear as an insoluble oil. Continue the heating, stirring the mixture occasionally until the oily layer disappears. Continue heating for another 30 minutes. Note observations on the Report Sheet.

Name _____ Partner _____

Section _____ Date _____ Due Date _____ Score _____

EXPERIMENT 40 ALKALINE HYDROLYSIS OF AN AMIDE: ACTION OF AQUEOUS SODIUM HYDROXIDE ON BENZAMIDE

1. Write the equation that symbolizes the observed behavior of benzamide toward aqueous sodium hydroxide. Below each condensed structure or formula write the name of the compound and state if it is water-soluble. (Benzamide is not soluble in water.)

Benzamide
(Insoluble)

2. Describe the odor emanating from the test tube after you have heated it. (Caution!)

3. How would the organic product present after completion of the reaction behave toward hydrochloric acid? Write a net ionic equation.

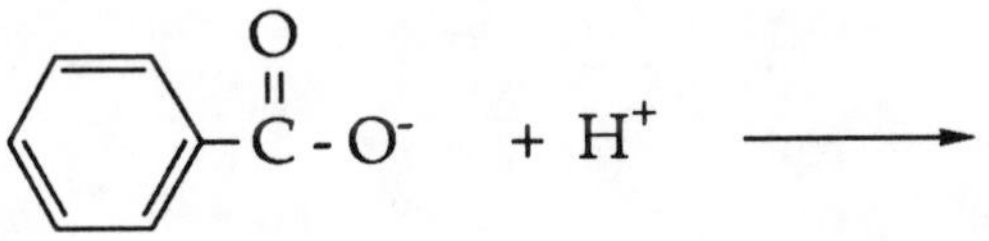

4. Cool the solution and add 1 mL of concentrated hydrochloric acid, note any changes, and explain them. Do not check the odor, because you will mostly sense the fumes of hydrochloric acid, which can be extremely irritating.

What did you see? _____ Explain: (write an equation)

EXPERIMENT 41 EXERCISES IN STRUCTURAL ORGANIC CHEMISTRY: CARBONYL COMPOUNDS

Distributed about the laboratory are a few ball-and-stick models of molecules of carbonyl compounds. Examine them, write neat condensed structural formulas in the space below, and the name of the compound. The number refers to the number on the model. Answer the question by each model on the blanks provided.

Structure and Name	Answers to Questions
1.	
2.	
3.	
4.	

5.

6.

7.

8.

9.

10.

CHAPTER 14   Carbohydrates

EXPERIMENT 42 STUDY OF CHIRALITY VIA MOLECULAR MODELS

Biochemistry, the study of the interactions between chemical substances in living systems, is highly dependent on nature of the molecules that comprise these systems.

Stereoisomers (space, or geometric isomers) are very important to us biologically, especially in the case of enzymes. They have the same molecular formulas and the same fundamental atom-to-atom sequence. However, because functional groups are arranged differently in a spatial or geometric sense, the body's enzymes can "fit" one isomer and use it, but not the other isomer. Some of these isomers are *cis*- and *trans*-isomers, which involve a double bond, or aliphatic rings that prevent free rotation and fix the orientation of the functional groups. Other stereoisomers are optical isomers which contain a tetrahedral stereocenter, or chiral center, a carbon to which are attached four different groups.

You have been studying isomerism ever since you started studying organic chemistry. In this exercise, we deal with *constitutional isomers*, which include isomers differing in their carbon chain connectivity, positional isomers (differing in the location of the same functional group), functional group isomers (differing in what functional group the atoms are arranged), and *stereoisomers*, which differ only in their geometry, especially *cis*- and *trans*-isomers, and optical isomers.

It is extremely helpful to view these types of isomers using molecular models, and for that reason, we have included Experiment 42.

This exercise can be done well by partners. The actual instructions for manipulating the models and the questions relating to the activities are found on the Report Sheet. Note that the following will be needed.

Procedure. Each partner should have the following parts of ball-and-stick models: 4 carbons, 10 hydrogens, 6 oxygens, 2 nitrogens, and 1 each of chlorine, bromine, and iodine; 10 short sticks (or "bonds"), 6 long sticks (or"bonds"), and 6 springs. Remember to be consistent: use long sticks between carbons, between carbon and oxygen, and between carbon and nitrogen; use short sticks between carbon and hydrogen, oxygen and hydrogen, or nitrogen and hydrogen. This allows uniformity in comparing formulas for superimposition, mirror images, and the like. Of course, use springs for double bonds.

EXPERIMENT 42 ISOMERISM VIA MOLECULAR MODELS

1. Chain isomers differ in their carbon skeletons. Each partner should make a model of butane. Are the two models identical? If so, convert one to an isomeric form of C_4H_{10}. Write the structural formulas of the isomers in the space at the right, and name them. Are more than two isomers possible?

2. Position isomers have the same molecular formula, but a substituent occurs in various positions, giving the two compounds different physical and chemical properties. Make two isomeric structures for butyl alcohol ($C_4H_{10}O$) in which there is a straight carbon chain. Write the structures here, and name them. Are more than two isomers possible?

3. Functional group isomers again have the same molecular formula, but because arrangements of the elements differ, the isomers belong to different classes of compounds. Write the structural formulas of ethyl alcohol and of dimethyl ether (C_2H_6O); name the formulas.

4. An important example of a pair of *cis-trans* **isomers** is fumaric acid (shown at right), which occurs in the citric acid cycle that you will study in carbohydrate metabolism. Its *cis*-isomer, maleic acid, is not usable by the body in this cycle. Construct models of both isomers, and write the structure of maleic acid.

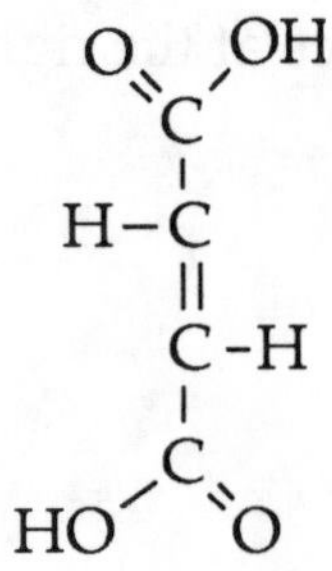

$$
\begin{array}{c}
O \diagdown \quad OH \\
C \\
| \\
H-C \\
\| \\
C-H \\
| \\
C \\
HO \diagup \quad O
\end{array}
$$

Fumeric acid

5. Construct models of methane, chloromethane, chlorobromomethane, and chlorobromoiodomethane. For which of these will the model of its mirror image NOT be superimposable? Make this mirror image and check it against its isomer.

Can you superimpose one above the other (any more than you can superimpose your hands, one palm on the back of the other hand!)? Now construct a mirror image of one of the other models. Try to superimpose this model on its mirror image. *Draw, and label, all your models below.*

6. Asparagine occurs in two optically active forms, one of which is shown on the next page. Make both the form shown and its mirror image. Can you superimpose the two models of asparagine, no matter how you twist the carbon chains? Draw the two models.

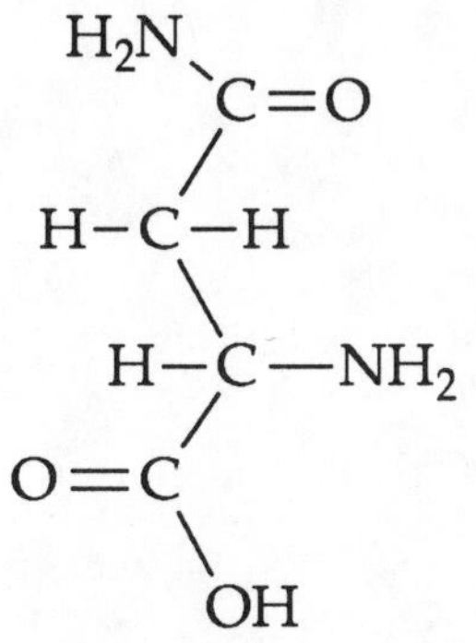

Asparagine

7. Optically active substances will rotate the plane of polarized light when in solution. However, in an equimolar mixture of the two enantiomers, called a *racemic mixture*, the rotation caused by molecules of one form is cancelled by molecules of the other form, and there will be no observable rotation. Make the mirror image of the tartaric acid shown at the left. Again, try to superimpose one above the other.

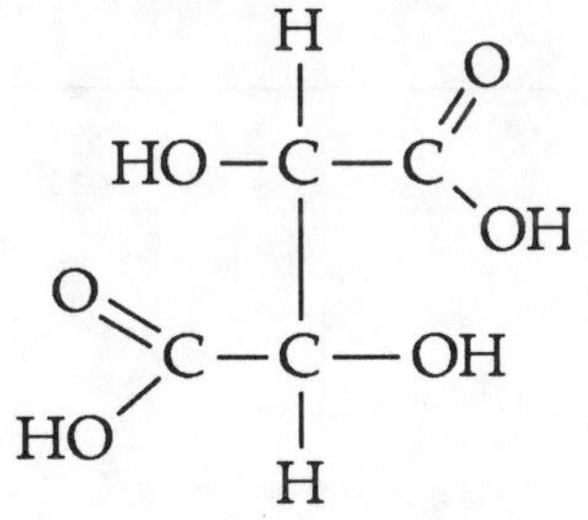

Tataric acid

9. Some pairs of models are distributed about the lab. Examine them. They may be structural isomers, constitutional isomers, functional group isomers, cis- and trans-isomers, or optical isomers. Write their structural formulas below, then tell which type of isomerism each pair represents. If they are *cis-* and *trans*-isomers, tell which formula is which. Name the compounds if you can.

Compounds A and B: Type of isomerism:_____

Structures:

Compounds C and D: Type of isomerism:_____

Structures:

Compounds E and F: Type of isomerism:_____

Compounds G and H: Type of isomerism:_____

Structures:

EXPERIMENT 43 TESTS FOR CARBOHYDRATES

The following tests identify specific characteristics of carbohydrates. Be sure to follow the directions carefully—if the reagents are to be layered, as in the Molisch test, do not mix them; if the reagents are to be mixed, as in the majority of the tests, mix them before putting them in the water bath. Be sure the water bath is at the required temperature before putting the tubes in the bath. If the meaning of a test depends on the rapidity with which a positive reaction is shown, as in the Barfoed and Seliwanoff tests, be sure you put all the solutions you may be testing into the boiling water bath at the same time, and write the time down on the Report Sheet.

You will also be given 15 mL of an unknown carbohydrate solution. Test this at the same time that you test the known solutions. Record the results on the Report Sheet, and by comparing these observations with those of the known solutions, identify the carbohydrate.

After each test, discard the wastes as directed by the instructor.

Molisch Test: *A general test for carbohydrates*

Procedure. Test each carbohydrate provided. Place 2 mL of each solution into a separate, clean test tube; label each test tube. To each test tube, add 2 drops of Molisch reagent (alpha naphthol in alcohol). Mix thoroughly. Incline the tube, and from the dropper provided, slowly and carefully add down the side of the tube 1 mL of concentrated sulfuric acid. It will run under the mixture of solution and reagent. Be extremely careful not to let any drops of this acid go anywhere but inside the tube. Keep the sulfuric acid dropper upright so acid does not run into the bulb. A purple color should develop at the interface between the solution and the sulfuric acid if carbohydrate-containing material is present.

Benedict's Test: *for reducing saccharides (presence of free or potential aldehyde)*

Procedure. Again, test each carbohydrate provided, allowing a separate, clean test tube for each. Label each tube. **Prepare a hot water bath.**

Pipet 3 mL of Benedict's solution into each test tube. To each test tube, add 5 drops of a carbohydrate solution, a different one in each tube. Mix each test tube by swirling.

Place all the tubes at the same time into the hot water bath (about half full of boiling water.) Since the solutions provided for this test will all be the same concentration, by noting which solution reacts first, you may get some clue about the nature of the sugar.

A positive test is the appearance of a red-orange precipitate of Cu_2O. However, the size of the particles may make it look orange or yellow—or even green if there is much of the deep blue cupric ion unreacted.

Optional: It is valuable to see how the amount of sugar present affects the color obtained. This optional experiment can be done as a group or by one partner while the other partner does the comparison of different sugars. Be sure to record the results of both experiments.

> Put 3 mL of Benedict's solution in each of five test tubes; to one tube, add 1 drop of glucose solution, and to the next tubes, add 2, 3, 4, and 20 drops. Put all solutions in the boiling water bath at the same time. Report your observations.

Barfoed's Test: *(distinguishes between monosaccharides and disaccharides).*

Procedure. Obtain a clean test tube for each carbohydrate to be tested. You will again need a **boiling water bath** for this test.

In separate test tubes mix 1 mL of each of the carbohydrate solutions to 3 mL of Barfoed's reagent. Mix, and place all tubes in a boiling water bath at the same time. Keep them in the boiling bath up to 10 minutes. Barfoed's reagent is copper acetate in acetic acid and not as reactive as Benedict's. A positive reaction may be only a little dark red precipitate in the bottom of the test tube. If this appears in 2 or 3 minutes, it indicates a monosaccharide; if it does not appear or does not show until about 10 minutes, a disaccharide is indicated. Keep records of the times required for evidence of reduction of the copper.

Seliwanoff Test: *for ketoses*

Warning: This reagent is resorcinol in concentrated hydrochloric acid, so handle carefully. Don't try to smell it—and don't let any of it go anywhere but in the test tube.

Procedure. A clean test tube for each carbohydrate solution is needed, in addition to a **boiling water bath**.

In each test tube place 3 mL of Seliwanoff's solution. Add to this, in each test tube, 3 drops of carbohydrate solution. Place all tubes in the water bath at the same time (note the time), and record the time again when a dark reddish-brown color appears in the clear solution (no precipitate). Compare the colors of all the tubes at the same time. A yellowish or apricot color is not a positive test. Do not heat for longer than this first change, because longer heating may give misleading results.

Bial's Test, *for pentoses.*

Warning. Be careful with this reagent! It contains concentrated HCl.

Procedure. Put 3 mL of Bial's reagent into a test tube and add 1 mL of sugar solution. Carefully heat the mixture in a flame until it boils; then let it cool. The appearance of a bright green color is a positive test for a pentose. If the color is not easily recognized, 1 mL of a higher alcohol such as pentanol may be added with shaking; the green color will be extracted by the alcohol if pentose is present. Fructose and sucrose interfere with this test since they produce a brown color.

Fermentation by Yeast

This test is negative for galactose and lactose. It is frequently doubtful in the case of maltose, which makes it difficult to distinguish between maltose and lactose. Yeast contains a great variety of enzymes and formerly contained enough maltase to hydrolyze maltose to glucose and give a positive fermentation test. However, the yeast developed for household use has been selected to be effective with sucrose and glucose, the sugars most used in cooking. The addition of a little maltase if possible to the solution will aid in the differentiation between maltose and lactose.

Procedure. If a solution of yeast is not provided for this experiment, make your own directly in your carbohydrate solution by weighing about 0.2 g of dry yeast or taking a very small piece of a yeast cake, and adding it to a very small test tube (about 4-mL capacity, or about 8 cm long). Have already prepared either a larger test tube into which the small one will fit, or a small beaker, 20- or 30-mL capacity. The test tube or the beaker should be about one-quarter full of solution. (If your solution is in short supply, a more dilute solution, or even water, can be used in the outside test tube or the beaker.) Also have ready a small circle of either filter paper or paper towel which will cover the end of the small test tube. Now about half fill your small test tube with the solution to be tested, close the end with a finger (clean), and shake well. When the solution and yeast are well mixed, fill the test tube to the brim with the solution and slip the paper circle across the top. It will be held by capillary action, and you can invert the tube and either slide it down the side of the larger test tube or upend it in the beaker of solution. Place the test tube in a water bath at about 40 °C; if a beaker was used, place it in a warm place (see Fig.29).

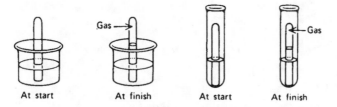

At start   At finish   At start   At finish

Fig. 29 Improvised fermentation tubes.

If a fermentable sugar is present, the most noticeable product will be carbon dioxide, which will rise as bubbles in the upper end of the test tube, displacing the solution. Examine the tube at the end of 15 minutes and again at 30 minutes if positive action was not noted earlier. The gas in an actively fermenting solution will keep the yeast stirred up. If the sugar will not ferment, the yeast will begin to settle.

Discussion. In Experiment 16 you tested for starch by adding a drop of iodine in potassium iodide reagent, obtaining a blue color. This color will disappear if the solution is warmed and reappear when the solution cools. It gets deeper near the freezing point of water. Glycogen gives a reddish color with iodine, but it is necessary to put the test tube in ice water to make it easily visible.

Name ______________________________ Partner ______________________________

Section __________ Date ___________ Due Date ____________ Score ______________

EXPERIMENT 43 TESTS FOR CARBOHYDRATES

UNKNOWN No. ___________

CARBOHYDRATE	MOLISCH	BENEDICT'S	BARFOED'S	SELIWANOFF	BIAL'S	YEAST	IODINE
GLUCOSE							
GALACTOSE							
FRUCTOSE							
(a pentose:)							
MALTOSE							
LACTOSE							
SUCROSE							
STARCH							
UNKNOWN							

My unknown was: ______________________

Questions:

1. You saw a purple ring at the interface in the Molisch test. Does that tell you which carbohydrate you might have? Explain.

2. Suppose you saw no sign of color change in Benedict's test, a dark red-brown solution with Seliwanoff test, and no sign of red precipitate after 10 minutes with Barfoed's test. The solution is showing signs of positive fermentation.

What is the sugar? ______________________

3. Suppose you saw an orange precipitate with Benedict's test, a definite dark red precipitate with Barfoed's test in about 2 minutes, a straw-colored solution after more than 5 minutes with Seliwanoff's test, and are getting signs of fermentation after about 15 minutes.

What sugar is present? ____________________

4. Suppose you saw all the same results as in number 3, except that there is no evidence of fermentation even after half an hour.

What is the sugar?_____

5. Suppose you saw an orange precipitate with Benedict's test, a straw-colored solution with Seliwanoff's test, and no evidence of precipitate with Barfoed's test in 10 minutes. The yeast is settling in the fermentation test; there is no evidence of gas formation. Can you narrow this down to one sugar, or do you have to say it can be one of two sugars?

What evidence would you need to distinguish between them?

EXPERIMENT 44 HYDROLYSIS OF SUCROSE AND OF STARCH

In this experiment, you will illustrate for yourself the results of hydrolysis of two carbohydrates, one a disaccharide and the other a polysaccharide. In both cases, the bond that is breaking through the action of heat and acid is the acetal between the ring structures.

Procedure. Place 5 mL of 1% sucrose in a test tube and add to it 2 drops of concentrated hydrochloric acid. Place 5 mL of 1% starch solution in another test tube and add 5 drops of concentrated hydrochloric acid. Heat the solution in each tube to boiling over a flame (carefully), then place the sugar-containing tube in a beaker of boiling water for 5 minutes, and the starch-containing tube in the boiling water for 15 minutes. Cool the solutions and neutralize each with 1 mL of 6 M NaOH. (Test with litmus paper.)

Procedure. Testing the resulting solutions for reducing sugar with Benedict's solution:

After you have heated the starch solution and the sucrose solution for the recommended time, take 5 drops of the starch solution and add it to 3 mL of Benedict's reagent in another test tube. Likewise add 5 drops of the sucrose solution to 3 mL of Benedict's solution in a clean test tube. Mix well. Place both test tubes in a boiling water bath. Heat for about 5-10 minutes or until a color change is observed. Note your results in questions 3 and 6 on the report sheet.

Procedure. Testing the starch hydrolysate with iodine:

The solution containing starch should be tested with iodine reagent. To do this, you should place a drop of the solution on a spot plate, and add a drop of the iodine solution, as in Experiment 43. A dark blue color indicates the presence of starch, and a red color, particularly if the solution has been chilled in an ice bath, indicates glycogen. If the color of the iodine reagent simply becomes diluted, no starch is present. Refer to question 7 on the report sheet.

EXPERIMENT 44 HYDROLYSIS OF SUCROSE AND OF STARCH

1. Draw the structure for sucrose, and circle the acetal link.

2. Does sucrose give a positive Benedict's test? (Is it a reducing sugar?) (Refer to data from Experiment 43.)

3. After treatment of sucrose with acid and heat, does the resulting solution give a positive Benedict's test? _____ Explain:

4. The presence of acid and HEAT were required for fast hydrolysis of sucrose. If a sucrose solution stands open to the air for a few days, it gives a test for reducing sugar. How can this be explained?

Write an equation, showing the atmospheric source of the acid:

5. Draw a cyclic structure for a simple starch, and circle one of the acetals.

6. After treatment of the starch solution with acid and heat, does the resulting solution give a positive Benedict's test?

Explain:

7. After treatment with acid and heat, does the solution give a positive iodine test?

Explain:

EXPERIMENT 45 EXERCISES IN STRUCTURAL BIOCHEMISTRY: CARBOHYDRATES

Distributed about the laboratory are ball-and-stick models of α-glucose, β-glucose, open-form glucose, β-maltose, β-galactose, fructose, sucrose, and possibly other sugars. Each model has a number. Below the model number on the line, write the name of its corresponding sugar.

Your instructor will demonstrate to small groups how two glucose molecules become joined together to form maltose; how ring-closure (cyclic hemiacetal formation) of the open-form of glucose can result in either α-glucose or, β-glucose; and how a developing amylose molecule differs from a developing cellulose molecule.

MODEL NUMBER:	NAME OF COMPOUND REPRESENTED:
1	
2	
3	
4	
5	
6	
7	
8	

Question: Describe ways in which the 3-dimensional models differ from the cyclic structural representations in your text.

CHAPTER 15

Lipids

EXPERIMENT 46 TESTS FOR LIPIDS

PHYSICAL PROPERTIES OF LIPID MATERIALS

46 A. Appearance.

Take a sheet of paper to the reagent table containing the display of lipids. Write the number of each, its chemical name and its common name if it has one. Then draw its full structural formula. Make brief comments on its physical properties. Feel the substances or smell them where instructed, but don't feel them or smell them unless instructed to do so.

46 B. Solubility.

Review the tests on solubility of polar and nonpolar substances you made in Experiment 30. Lipids are generally very nonpolar. The only lipid class easily soluble in water are the phospholipids, which will be the subject of Experiment 47.

46 C. Spot Test for Fats and Fatty Acids.

From mishaps when cooking or eating, most of us are all too familiar with the spot test for fats. To make it "official," however, take a piece of filter paper and mark on the edge with pencil "stearic acid," "oleic acid," "vegetable oil," "soap," "syndet, " and any other lipid type of substances you wish to include. Apply a drop of the solution of the designated substance to the paper opposite its name, using the pipet from the solution. Place the paper on a small beaker to let the spots dry. Hold it up to the light when dry. Are there any translucent spots, or just stains? The area of the paper where the stearic acid and soap were deposited can be held near a flame or hot light bulb or placed on a flat glass plate over a beaker of boiling water—to see whether a translucent spot appears. Record the appearance of the paper for each substance.

CHEMICAL PROPERTIES OF LIPID MATERIALS

46 D. Unsaturation of Lipids.

Review the use of Br_2 in methylene chloride to detect unsaturation (Experiment 28). **Use dry test tubes.**

Procedure. Measure 2 mL of methylene chloride solution of stearic acid and of at least two glycerides (one animal and one vegetable fat) into separate test tubes. Take them to the hood, and, above a white background carefully add 1 drop of Br_2 in methylene chloride solution to the stearic acid solution. If it does not show a yellow tint, add one more drop. It should turn yellow. Now add Br_2 in methylene chloride dropwise to the glyceride solutions, up to 10 drops. Does either one turn yellow? Record the results on the Report Sheet. Dispose of the test solutions in the designated container.

46 E. Test for Glycerol.

Procedure. Mix 1 drop of glycerol with a very small amount (1-1.5 g) of powdered potassium hydrogen sulfate in a dry test tube and heat it gently over a flame, in a hood. Move the test tube continually, shaking it from side to side with a slight up-and-down rotary motion, or put a stirring rod in it and stir carefully while you are heating. The mix should melt slightly and blacken slightly, and cloudy fumes should start to rise from within the tube. Remove the tube from the heat and smell it very carefully (see Fig. 30). The glycerol is dehydrated and otherwise changed to acrolein (see question 5 on Report Sheet). This has a very noxious odor reminiscent of poorly ventilated hamburger stands.

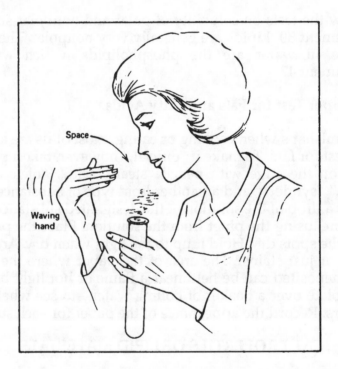

Fig. 30 Technique for detecting odors.

Repeat the experiment using stearic acid instead of glycerol and note that there is not much melting, blackening, or odor. Now repeat the experiment using about 3 or 4 drops of the vegetable fat or oil, and then using a similar amount of a solid fat. Do not use a solution of fat or oil. If you do not heat the test tubes so much that the residues become hard and dry, they will wash well with solid detergent and a brush.

46 F. Liebermann-Burchard Test for Cholesterol.

Procedure. The test tubes must be dry. Measure 2 mL of cholesterol in methylene chloride solution into a dry tube, and 2 mL of each of two glyceride solutions into two other dry test tubes. Add to each 10 to 12 drops of fresh acetic anhydride; mix carefully. Add 3 drops of concentrated sulfuric acid (care), and again mix carefully. Watch for color changes. A blue-green color is characteristic of cholesterol. If no green color appears immediately, let the mixture stand up to 10 or 15 minutes before discarding. Other colors may appear earlier, but they must revert to green to be a positive test for cholesterol. Discard solutions in the designated container. If a fat is rancid it may produce muddy colors that will interfere with the cholesterol test. If that happens, record such a result. Compare your observations with those obtained by other class members.

Name ______________________________ Partner ______________________________

Section __________ Date __________ Due Date ____________ Score ______________

EXPERIMENT 46 TESTS FOR LIPIDS

46 A. Observations

1. Name:	4. Name:
2. Name:	5. Name:
3. Name:	6. Name:

46 B. List those lipids that you found soluble in water:

List those lipids that you found soluble in ether or another nonpolar solvent:

46 C. It is suggested that the paper with the spot tests be stapled to this report. List the substances on your filter paper that left translucent spots on drying:

List the substances that showed translucent spots when warmed (the melting point of stearic acid is about 70 °C):

46 D. Which fats showed unsaturation when you tested them with bromine?

Did you find any difference in the amount of unsaturation of the animal and vegetable fats? (If there are several substances that can be tested, find out what results your neighbors obtained on substances you didn't test.)

46 E. Observations:

Some packaged dessert mixes contain "mono- and diacylglycerides." Would you get a test for glycerol for these, as well as for triacylglycerides? Why or why not?

46 F. Which fats showed strong Liebermann-Burchard test for cholesterol?

Which fats showed a weak test for cholesterol?

Which fats showed no test, or showed interference?

Define "steroid."_____

_____

Draw the structural formula for one steroid substance and label it:

EXPERIMENT 47 PROPERTIES AND CONSTITUENTS OF A PHOSPHOLIPID (Lecithin)

Review the subject of phospholipids in your text. On the Report Sheet, draw the structure of a molecule of the lecithin class. Esterify an unsaturated acid on the middle OH of the glycerol. Mark with arrows all the places in the molecule where hydrolysis (or saponification) can occur. Then draw the structures that would result from complete hydrolysis of the molecule, and name all of these products.

Lecithin is a very common representative of the glycerophospholipids, which are the most polar of the lipids. They are found chiefly in cell membranes, where they have important functions in transport because of their combination of nonpolar fatty acid chains and polar or ionic groups elsewhere in the molecule. The presence of both polar and nonpolar groups greatly affects the solubility behavior of phospholipids compared with true triglycerides (see Experiment 46). Lecithin is ordinarily a soft, waxy solid that darkens on exposure to air, since at least one of the fatty acids present is unsaturated. The lecithin preparations sold in natural food stores are combined with related substances and frequently are dry granules so as to be more easily handled. These preparations do not pretend to be pure phospholipid and contain other substances associated with cell membranes, including inositol and cholesterol.

47 A. Solubility of Lecithin in Water.

Procedure. Dissolve a very small amount of lecithin in 5 or 6 mL of water. It may take some "work" in the form of stirring. Lecithin will not form a clear solution, but rather a cloudy emulsion; this will not separate, however. Compare this behavior with that of 2 or 3 drops of vegetable oil shaken with about 3 mL of water. Now take half of your lecithin emulsion and add 2 or 3 drops of vegetable oil; shake. Notice whether the oil separates from the lecithin emulsion as it did from the water. This demonstrates the emulsifying power of lecithin and explains why lecithin is found among the ingredients of so many products such as cake mixes, candies, and the like.

47 B. Solubility of Lecithin in Nonpolar Solvents.

Procedure. Try to dissolve a small amount of lecithin in 4 or 5 mL of methylene chloride. Is it soluble? (Again, stirring is needed.) Divide this solution into two dry test tubes for use in the next two tests.

47 C. Presence of Cholesterol.

Apply the Liebermann-Burchard test (Experiment 46).

47 D. Presence of Unsaturation.

See Experiment 28. If your solution already has a yellow cast, take an equal volume of methylene chloride in another dry test tube and add bromine in methylene chloride drop-by-drop (count) to the methylene chloride until you have obtained the same tint as your lecithin solution. Now add the same number of drops of bromine in methylene chloride to your lecithin solution. Does it change color? Can you add more bromine than this without changing the color of the lecithin solution?

47 E. Presence of Glycerol.

Apply the acrolein test of Experiment 46.

47 F. Presence of Amine Group and of Fatty Acids.

If the lecithin is heated with alkali, its ester links between glycerol and fatty acids are saponified—at the same time the alkali will free the ammonia from an amine group.

Procedure. Into a 50-mL Erlenmeyer flask or a wide test tube (about 25 x 150 mm), place 7 or 8 mL of 20% KOH solution and add an equal volume of isopropyl alcohol. Then add 1 g of lecithin. (If it is a waxy preparation, weigh a small stainless steel spatula, then add lecithin to approximately 1 g and transfer the lecithin directly to the alkali, stirring it in as you heat it; see below.) Heat this mixture in a water bath at about 75 °C, (no hotter). Isopropyl alcohol boils at 82 °C, and its vapors catch fire easily from the burner. Swirl or stir the mixture frequently. When the lecithin is saponified, an emulsion forms; continue to heat it another 5 minutes (about 10 minutes total). While heating is in progress, hold a piece of moist red litmus in the neck of the flask or test tube and note the blue color produced, indicating the production of ammonia from the amine present.

Dilute this saponification mixture with an equal volume of water; a clear solution should result. Acidify this solution with concentrated hydrochloric acid, a few drops at a time (care) with stirring. Note the "smoke" formed when the acid is added the ammonia formed from the amine is reacting with HCl to form solid ammonium chloride. When the solution stays cloudy after stirring, test it for acidity by touching the stirring rod to a piece of litmus paper. Add concentrated HCl until the solution tests acid. The liberated fatty acids have now separated from the solution. (Why?) Immerse the mixture in an ice bath for a few minutes, then filter it through a small filter paper wet with ice water. The fatty acids from the original lecithin will be retained by the cold wet filter paper. Note their color. Their solubility in alkali can be tested, and the fact that they are unsaturated can also be shown again at this point if desired.

47 G. Presence of Phosphate.

The carboxylate-ester bonds can be easily saponified, but not the phosphate-ester bonds (those between glycerol or choline and phosphate). These must be hydrolyzed with acid in order to free phosphate for detecting with molybdate.

Procedure. To about 0.2 g of lecithin in a test tube add 3 mL of 6 M nitric acid, and place it in a boiling water bath for 5 or 6 minutes. There will be some frothing which may have to be stirred down. Do not inhale these vapors! A somewhat waxy-looking solid will be floating on the surface of the mixture. Place the tube in an ice bath for 5 or 10 minutes, then decant the solution through a very small filter wet with ice water. Add 1 or 2 mL of ammonium molybdate solution (see Experiment 20) and mix, then place the tube in a water bath no warmer than 65 °C for a few minutes. A precipitate of yellow ammonium phospho-molybdate should form in a few minutes.

Name ______________________________ Partner ______________________________

Section __________ Date __________ Due Date __________ Score ______________

EXPERIMENT 47 PROPERTIES AND CONSTITUENTS OF A PHOSPHOLIPID (Lecithin)

PRELAB EXERCISE: Structural formula of a lecithin molecule:

1. Did the lecithin form an emulsion in water? Did it separate?

Did this emulsion hold added vegetable oil so the oil did not separate?

Compare with vegetable oil-water mixture:

2. Was lecithin soluble in nonpolar methylene chloride?____________

3. Was cholesterol present? _________Much or little?_____________

4. Was unsaturation present?__________Much or little?___________

5. Glycerol present?_________________

6. Amine group?___________________

Write the structural formula for one of the fatty acids in your lecithin formula as it would appear after saponification.

Write the structural formula for one of the fatty acids in your lecithin formula as it would appear after acidifying with HCl.

7. Did you show the presence of phosphate in the phospholipid?

EXPERIMENT 48 SAPONIFICATION: THE PREPARATION OF SOAP; PROPERTIES OF SOAP COMPARED TO SYNTHETIC DETERGENTS

48 A. Preparation of Soap

Precautionary Measures

You will prepare soap by boiling a small sample of a typical vegetable oil with 20% sodium hydroxide. To conserve time, the mixture must be boiled as strongly as possible until the water is largely expelled. Caution: This step can be dangerous because the mixture spatters easily. As always be sure that your safety glasses are in place. The experiment can be performed safely if the following precautions are observed:

1. Regulate the rate of boiling and stir the mixture vigorously with an 8-in rod while holding the beaker with tongs. This keeps the hand far from the top of the beaker. If disposable gloves are available, wear them. Be prepared to withdraw the burner quickly, especially as the reaction nears completion. Although the boiling action must be vigorous, as the mixture thickens, there is the danger that charring may occur.

2. Before applying heat to the mixture, check to be sure that the stirring rod may be left in the beaker without tipping it over. You may wish to interrupt stirring to rest your arm.

3. Keep your head well back from the beaker. In spite of precautions spattering may yet occur.

4. When you pause to rest, remove the burner, but continue the stirring until boiling action ceases.

5. When you resume heating, start stirring the mixture before you begin to apply heat.

Procedure. Use beaker tongs, be sure to have eye protection. Prepare a mixture of 15 mL of 20% (5 M) sodium hydroxide and 10 mL of vegetable oil in your (100- or 150-mL) beaker. Boil and stir this mixture observing all the precautions. Be especially careful to control the burner near the end of the reaction to prevent charring. When it appears that all the water has been expelled, allow the mixture to cool slightly. The saponification is complete if a waxlike solid begins to form that on further cooling becomes hard and somewhat brittle. On the other hand, if the mixture cools to a syrupy liquid, saponification is not complete, and heating and stirring must be resumed. It might be advisable to add more (5 mL) 20% sodium hydroxide and boil the mixture until its water is expelled.

When saponification is complete, allow the mixture to cool to room temperature. **Do not handle the product yet because it still contains considerable amounts of sodium hydroxide ("lye").** It also contains glycerol.

Both the glycerol and sodium hydroxide may be washed away as follows:

Washing Procedure. While your crude soap cools, prepare a concentrated solution of sodium chloride by dissolving 18 g of sodium chloride in 60 mL of distilled water. Using 20-mL portions of this solution each time, wash the solid in the beaker by stirring it with the salt solution. Using a spatula, break up the lumps as completely as possible to permit maximum contact between the solid and the wash solution. Decant the wash solution from the soap. (If the soap tends to float, you may prevent it from leaving the beaker with the wash solution by holding your wire screen over the top of the beaker. If you do this, be sure to scrub off the soap as soon as possible; before again heating it!) Repeat this operation two more times with the salt solution. After the final washing, remove the last traces of liquid by working the soap on a piece of towel or filter paper.

There are no responses on the Report Sheet for 48 A. You may use the soap that you made for second part of this experiment (48 B).

49 B. Comparison of the Properties of Soaps and Detergents

Specific instructions and questions will be found directly on the Report Sheet.

EXPERIMENT 48 SAPONIFICATION: THE PREPARATION OF SOAP; PROPERTIES OF SOAP COMPARED TO SYNTHETIC DETERGENTS

48 B. PROPERTIES OF SOAP COMPARED TO SYNTHETIC DETERGENTS

Note: Procedures are noted in the usual typeface. Look for them.

Properties of Soap

Construct a structural formula of a lipid molecule that incorporates the following carboxylic acids ("fatty acids") in its structure: palmitic acid, linoleic acid, and oleic acid (consult your textbook). Using this structural formula, write a balanced equation symbolizing its saponification.

Alkalinity. Dissolve a small piece of your soap (about the size of a pea) in 5 mL of distilled water. Add 3 to 4 drops of phenolphthalein indicator solution. What color (if any) appears? What does it signify about the acidity or basicity of the solution?

Repeat the experiment with a commercial soap of known high purity (e.g., Ivory). What color did phenolphthalein produce? What does it indicate about the acidity or basicity of the solution?

In the case of your soap preparation, it is entirely possible that you did not wash out all the unreacted sodium hydroxide. However, the behavior of the commercial soap indicates that some other factor may be operating. To discover what that might be, write ionic equations that show how the sodium

Lipids

salt of any fatty acid would hydrolyze in water. (Remember that fatty acids are weak acids and that sodium hydroxide is a strong base.)

What conclusion may be drawn about the possibility of a solution of pure soap in distilled water having a pH of 7?

Lathering Qualities. Wash your hands with a small sample of your soap. Does it have satisfactory cleansing qualities?

Behavior Toward Hard Water. Place 5 mL of distilled water in one test tube, 5 mL of tap water in a second, and 5 mL of 1% calcium chloride in a third. Add small (about the size of a pea), equal quantities of your soap to each and shake the tubes very vigorously. Describe the relative extents to which lather and foam appear in each tube.

Distilled water:_____

Tap water:_____

1% calcium chloride:_____

Write a balanced net ionic equation that symbolizes the behavior of calcium ions toward carboxylate ions (negative ions from fatty acids).

Behavior Toward Salt Water (e.g., Seawater). Dissolve a small portion of your soap in a minimum amount of distilled water at room temperature Add an equal volume of concentrated sodium chloride solution (which you can prepare yourself). Describe what happens.

Explain why ordinary soap is rather ineffective when seawater (as on board some ocean-going vessels) must be used for washing purposes.

Explain why concentrated salt solution was used to wash your crude soap preparation rather than distilled water (or tap water).

Soap is the sodium salt of a fatty acid. You showed it to be soluble when you prepared the soap solution. To show why it behaved as it did with excess sodium ion, review Experiment 13, then write ionic equations showing how excess sodium ion will shift the equilibrium and force sodium stearate out of solution.

Properties of Synthetic Detergents ("Syndet").

Now write the structure of lauryl alcohol ($C_{12}H_{25}OH$) and the equation for its conversion to an ester with sulfuric acid. Then write another equation showing the action of sodium hydroxide on this ester, resulting in the formation of sodium lauryl sulfate, a prototype of the first synthetic detergent. Note its hydrocarbon "tail" and hydrophilic "head."

Place 2 or 3 mL of synthetic detergent solution in each of three test tubes. Add 5 drops of calcium chloride solution and shake the tube. Describe what happens:

Why is a synthetic detergent superior to soap for sudsing in hard water?

To 3 mL of detergent solution add 5 drops of sodium chloride solution and shake the tubes. Describe what happens:

Syndet + NaCl: Repeat the above experiment with "synthetic " sea water, if it is available. This will contain sodium, calcium, and magnesium ions in the proportions found in sea water. Describe what happens.

Phosphates. Test your soap solutions and synthetic detergent for phosphate by adding 5 drops of dilute nitric acid and 2 mL of ammonium molybdate to 2 mL of your solution. Warm in water bath or in the flame, but do not boil. A yellow precipitate is a positive test for phosphate. Compare your synthetic detergent with others brought to the class.

Detergent (name)	Phosphate Present?
_____	_____
_____	_____
_____	_____
_____	_____

CHAPTER 16

Proteins

EXPERIMENT 49 REACTIONS OF PROTEINS

PART 1: COLOR REACTIONS

A variety of tests exists to determine whether a substance or a solution contains proteins or protein-breakdown products (e.g. proteoses, peptones, polypeptides, and in some tests, amino acids). These tests depend on the fact that certain functional groups present in these substances will react with a specific reagent to produce a change in odor, color, and/or solubility. (See your text for more details.) What is interesting here is the fact that the side chains of these (often very large) peptides, as well as their amide "backbone" act chemically as if they were independent of the rest of the molecule. This allows us to test for the presence of particular amino acid residues. Certain types of proteins have an abundance of some amino acids and a deficiency of others. This becomes important in nutrition. It also has implications for the characteristics of certain proteins: water-solubility or insolubility, for example.

Biuret Test. This is a general test for protein comparable to the Molisch test for carbohydrate.

Reagent: A very *pale blue* solution of copper sulfate (0.1%) is used following action of sodium hydroxide.

Group Detected by the Reagent. A violet color appears when this reagent combination is added to any compound (protein or otherwise) which contains two or more of the following groups:

$$\overset{\displaystyle \overset{O}{\|}}{-C-NH_2}$$

as in
Asparagine

$$\overset{\displaystyle \overset{O}{\|}}{-C-NH-}$$

Amide
(peptide backbone)

$$\overset{\displaystyle \overset{NH}{\|}}{-C-NH_2}$$

as in
Arginine

$$-\underset{\underset{OH}{|}}{CH}-CH_2-NH_2$$

as in
Threonine

$$-CH_2-NH_2$$

as in
Lysine

The name of the test is derived from a specific compound, *biuret*, which gives the test.

$$NH_2 - \overset{\overset{\displaystyle O}{\|}}{C} - NH - \overset{\overset{\displaystyle O}{\|}}{C} - NH_2$$

Biuret

Procedure. Thoroughly mix 2 mL of 3 M sodium hydroxide with 2 mL of the solution to be tested. Add 1 drop of 0.1% copper sulfate solution. Mix thoroughly and note if a pink or violet color develops. If not, add additional drops (up to 10) of 0.1% copper sulfate, mixing the solution after each addition. A positive test is the appearance of a pink or violet-blue color; it may be very pale. Use a white background. Omit phenol.

Xanthoproteic Acid Test. A test for the presence of tryptophan, phenylalanine. tyrosine, serine, and threonine.

Reagent: Concentrated nitric acid. *(Care! Keep from clothes or person.)*

Groups Detected by the Reagent. Benzene rings on which there are amino groups (e.g. tryptophan) or hydroxyl groups (e.g., tyrosine) are easily nitrated to give yellow-colored aromatic nitro compounds, which are given the general name, xanthoproteic acid (*xantho*, Greek word for yellow)

Procedure. Add 1 mL of concentrated nitric acid to 2 mL of the solution to be tested. Mix and note the appearance of any heavy white precipitate. Warm the mixture carefully, noting any change to a yellow-colored solution. Cool the mixture in a stream of cold water and carefully add 3 M sodium hydroxide and note if the color deepens to orange. This is a positive test. The entire tube does not have to turn orange. Look for the color where the sodium hydroxide has entered the test solution, or on pieces of precipitate on the wall of the tube. Try all four substances.

Hopkins-Cole Test for the presence of tryptophan.

Reagent: Hopkins-Cole reagent is a solution of magnesium in oxalic acid. It contains glyoxylic acid:

$$H - \overset{\overset{\displaystyle O}{\|}}{C} - \overset{\overset{\displaystyle O}{\|}}{C} - OH$$

Glyoxylic acid

Groups Detected by the Reagent. The indole of the tryptophan system is specifically detected by the reagent. The chemistry of the test is not completely understood. Follow directions *very* carefully.

Procedure. Add 2 mL of Hopkins-Cole reagent to 2 mL of the solution to be tested in a test tube. Incline the tube at an angle and *carefully* add 3 mL of concentrated sulfuric acid so that it drains down one side and forms a separate *lower* layer in the tube. A positive test consists of the appearance of a violet ring at the interface in a few seconds. If this fails to appear, very *gently* agitate the tube to provide a *slight* mixing at the interface without bringing about extensive solution of the two layers in each other. Omit phenol if instructor so directs.

IDENTIFICATION OF AN UNKNOWN PROTEIN

You will be given a test tube containing at least 15 mL of one of the three proteins you have tested in the preceding experiments. If you review your Report Sheets for these experiments, you should be able to make your identification very quickly.

PART 2: DENATURATION OF PROTEINS

You will need about 12 mL of each of the protein solutions for these tests. Record your observations on the Report Sheet. **Procedures:**

Action of Heat. Heat the *upper* portion of 5 mL of each solution except gelatin, until it boils. Compare it with the lower section. This test is sometimes used to identify albumin in urine.

Action of Alcohol. Add 1 mL of denatured alcohol to 1 mL of protein solution. Note what you see.

Action of Heavy Metal Ions. Add several drops of mercuric chloride solution to 1 mL of each protein solution. Repeat, using fresh solution, with silver nitrate. Repeat, using fresh solution, with lead nitrate or lead acetate. Note what you see.

EXPERIMENT 49 PART 1 COLOR REACTIONS OF PROTEINS

When filling in the outline below, add information that will help you use these tests to identify an unknown protein.

	Biuret	Xanthoproteic	Hopkins-Cole
Albumin			
Casein			
Gelatin			
Phenol			
Unknown # _____			

EXPERIMENT 49 PART 2 DENATURATION OF PROTEINS

Denaturing Agent:	Albumin	Casein	Gelatin
Heat			
Alcohol			
Heavy metals: mercury ions			
silver ions			
lead ions			

Questions:

1. Compare the solubility of the three proteins in hot water:

2. Why has alcohol or mercuric chloride been successful as a disinfectant?

3. What amino acids in a protein are particularly reactive toward heavy metals?

4. People who have swallowed such poisons as mercuric salts, copper salts, or lead salts are given egg white or milk. Why?

Such treatment must be followed by an emetic. Why?

5. What structural change occurs in a protein molecule when it is denatured?

EXPERIMENT 50 pH AND PROTEIN SOLUBILITY: ISOELECTRIC POINT

Note: there are PRELAB questions (50 A.) that should be done before coming to the lab. Refer to the Report Sheet.

Proteins are classified in several ways but most commonly by *water solubility* or *by function*. Actually, these two are related, for a protein that is soluble would be useless for tendons, fingernails, or hair. An insoluble protein would not make a very good hormone or enzyme or blood buffer. As you showed in Experiment 49, water-soluble proteins are rendered insoluble by certain denaturing agents. pH as a denaturing agent is considered in this experiment.

If done carefully, the following test can be used in conjunction with the color tests of Experiment 49, for differentiation between albumin, casein, and gelatin.

51 B. Procedure. Do these tests slowly and carefully. Place 2 mL of the solution to be tested in a clean test tube and add 2 drops of universal indicator solution. Add 0.05 M sodium hydroxide drop by drop, counting drops, noting changes in color of indicator, or in clarity of solution. Add another drop of indicator for each 10 drops of base added. Especially note whether, at any particular pH, several (how many?) drops can be added without changing the pH. This will indicate buffering, and represent an isoelectric point. Continue to add sodium hydroxide drop by drop until pH 9, *or* the highest pH of the indicator color range, is reached. If you reach an isoelectric point, count the drops of base you can add before the pH again begins to increase. After you pass the isoelectric point, does each drop of base cause a big change in the pH of the solution?

Take a fresh 2-mL sample of the protein solution, add 2 drops of indicator, and add 0.05 M acetic acid drop by drop, observing the pH with each drop of added acid. Continue to add the acid until there is indication of an isoelectric point (clouding, formation of a precipitate, no change in pH). Does any precipitate formed redissolve when more acid is added? Add a drop of indicator for each 10 drops of acid added. Do not add acid after the first drop that brings the solution to pH 4, or the lowest pH in the color range of the indicator solution. Albumin will show some buffering at the isoelectric point but may not show clouding if it is too dilute. Carefully warm the tubes with acid solution to near boiling. Is the protein soluble in hot acid solution?

EXPERIMENT 50 pH AND PROTEIN SOLUBILITY: ISOELECTRIC POINT, pI

50 A. Before Coming to the Laboratory

1. In the space below, write the structure of a short polypeptide that contains at least one amino acid with an acidic side chain and one with a basic side chain.

   (a) Show how the solubility of such a protein would be affected by a change in pH by writing the structure as it would appear in an excess of acid,

   (b) and in an excess of base.

   (c) Show the structure at its isoelectric point.

   (d) For the structure you have drawn, would the isoelectric point (pI) occur at acid, basic, or neutral pH?

2. When you add acid or base to a protein solution, what evidence will you be looking for that will tell whether or not you are at or near the isoelectric point?

3. Why is a protein more soluble when it is bearing excess (+) or (-) charges?

4. Define "isoelectric point" in terms of charge.

50 B. Experimental

	Albumin	Casein	Gelatin
pH of protein solution before adding acid or base:			

Testing the BASE range of pH for an isoelectric point: *addition of base*

	Albumin	Casein	Gelatin
1. Total drops added to first reach pH 9 (blue color).			
2. Did the solution become cloudy, or did precipitate form at any time?			
3. Did you add drops of base with no change in color (pH) (indicating a buffer region)? If so, record the number drops added in this region.			
4. pH if an isoelectric point is observed.			
5. Was the protein soluble in basic solution (i.e. on the basic side of the isoelectric point)?			

Testing the ACID range of pH for isoelectric point: *addition of acid*

	Albumin	Casein	Gelatin
1. Total drops added to first reach pH 4 (orange color).			
2. Did the solution become cloudy, or did a precipitate form at any time?			
3. Did you add drops of acid with no change in color (pH) (indicating a buffer region)? If so, record the number of drops added in this region.			
4. pH if an isoelectric point is observed:			
5. Was the protein soluble on the acid side of the isoelectric point?			
6. Is the protein soluble in acid when it is heated?			

Questions:

1. Compare the solubilities of casein and of albumin in acid. These proteins both contain side chains with free amine and free carboxylic acid groups. Which of these groups produces solubility in acid?

Which of these two proteins appears to have a higher ratio of amine groups to carboxylic acid groups, based on solubilities?

2. Look up the structural formulas of the amino acids and list those that have amine-containing side chains which can contribute to solubility:

3. Compare the buffering ability of the three proteins:

4. Would the number of drops of acid required to go from pH 7 to pH 4 vary with concentration of the protein? (Be specific)

With which protein would it make the least difference, and why?

5. Is the albumin in blood (pH 7.25) in the protonated form, or does it bear excess (-) charges?

If you were preparing a solution to be transfused into blood would you need to pay attention to the pH as well as to the concentrations of dissolved particles? Explain.

Gelatin contains a high proportion of a relatively uncommon amino acid, hydroxyproline. Study its structure. The amine group and the acid group are both incorporated into the peptide bonds on either side in the polypeptide. Would this amino acid affect gelatin's solubility in acid or base? How would it affect solubility? Explain:

Hydroxyproline

EXPERIMENT 51 EXERCISES IN STRUCTURAL BIOCHEMISTRY: LIPIDS, AMINO ACIDS, AND PEPTIDES

Distributed about the laboratory are ball-and-stick models of lipid and lipid-related molecules, amino acids, and di- and tripeptides. Examine them. Write neat structural formulas and the name of the substance in the spaces provided, and answer the number-coded question by each model.

Structure and Name	Answers to Questions
1.	
2.	
3.	
4.	

5.

6.

7.

8.

9.

10.

CHAPTER 17   Enzymes

EXPERIMENT 52 FACTORS AFFECTING ENZYMATIC ACTIVITY

Work in pairs for this experiment. Read and study each part before you begin. Certain operations must be carried out quickly, and you must have your equipment and chemicals ready, and know what to do.

The digestion of starch is catalyzed by the enzyme amylase, which is found in saliva. Chloride ions are known to be activators and as with all enzymes, amylase will have highest activity at a particular pH and temperature (or ranges of these). Amylase is a protein and is therefore susceptible to denaturation by a variety of agents and conditions, some of which will be studied in this experiment. In part A you will study how temperature affects the activity of amylase. In part B the influence of pH on the salivary digestion of starch will be examined, and in part C you will study what a heavy metal can do to enzymic activity. Part D investigates the effect of the concentration of the enzyme on the rate of digestion.

The amylase needed for these experiments will be contributed by the experimenters, and this introduces a problem. There are widely differing amylase activities in the salivas of different people. Consequently some of you will obtain disappointing, inconclusive results in the search for evidence that digestion of starch is occurring. Others may discover amylase activity so high that it is difficult to stop it. A group discussion following the experiment would be a good way of comparing results. Another way to "even out" this variation is to combine the salivas of two or more people and use the mixture. The pancreatin for part D is provided.

Assemble the following equipment:
- a thermometer (2, if available)
- 5 or 6 dropper pipets
- 3 large (400-800 mL) beakers
- 2 small (100-250 mL) beakers
- 10-12 large test tubes
- a porcelain spot plate

Preliminary Work

1. Prepare constant temperature baths: You will need four of these. Fill the three larger beakers half to two-thirds full of water, and heat one to 37 °C, one to 70 °C and one to boiling. Once these have come to the proper temperature, maintain that temperature as closely as possible, adding water as necessary to replace water lost by evaporation. The fourth beaker

(150 or 250 mL) should contain an ice-water slush, at 0 °C.

2. Collect saliva: You and your partner will have to contribute 2 mL each of saliva, collected in a small, clean, dry beaker, or a large test tube. (Daydream a little about your favorite food while making your contributions.) When the saliva collection is complete, combine the contributions in one small beaker, mix well, and save this mixture for parts A, B, and C of the following experiments.

3. Obtain iodine solution: You will need a few milliliters of the dilute iodine test solution either in a dropper bottle or in a test tube with a dropper kept in the solution, ready for use. This should be kept at your work station. When you use this solution, four drops of the starch-saliva mixture should be placed in a depression on the clean porcelain spot plate, with 1 drop of the iodine solution. If starch is present, you will see a dark blue color. If there is no change from the amber color of the reagent, then starch has been digested and is no longer present as such.

4. Obtain the Benedict's solution: You will need 15-20 mL of Benedict's reagent. In each of 5-6 test tubes, place 3 mL of the blue reagent. Have these ready for Part A. When you use this, a test tube containing the 3 mL of Benedict's reagent plus four drops of the saliva-starch mixture you are testing must be placed in the boiling water bath for at least 5 minutes (often longer). If a reducing sugar is present (indicating that the starch has been digested), the blue color of the reagent will change to green or orange or brick-red, depending on the amount of sugar present. Again, no color change means that no digestion (or incomplete digestion) has taken place.

52 A. THE INFLUENCE OF TEMPERATURE ON ENZYME ACTIVITY

Procedure.

1. Put 5 mL of a 1% buffered starch solution in each of three test tubes. Place one tube in the 37 °C water bath, one in the 70 °C bath, and one in the ice-water bath. Leave them in their respective baths, occasionally stirring or swirling them, until they have reached the temperature of the bath.

2. From your saliva "stock", prepare a dilute solution by mixing 1 mL of saliva with 50 mL of distilled water. Mix the solution well, then transfer 5 mL portions into three more clean test tubes. Place one in each of the temperature baths as in step one. You now have a tube of starch and a tube of diluted saliva in each bath (37 °, 70 °, and 0 °C). Allow them to come to the temperature of the bath. Save the rest of the diluted saliva mixture for part C.

3. Begin with the test tubes in the 37 °C water bath.

    (a) Remove the two test tubes, and pour the contents of one into the other, mixing quickly and thoroughly. With the starch-and-saliva mixture all in one test tube, return the tube to the bath.

    (b) Immediately withdraw 4 drops of the mixture and put it on the spot plate. Test it with a drop of iodine solution, and record the color.

(c) Immediately withdraw another 4 drops and add this to one of the test tubes containing the Benedict's reagent. Mix, and place in the boiling water bath. (These are the "0" minute tests.)

(d) Note the time. You will be repeating the routine above (items b and c) every three minutes for the next 12 minutes, and after that, every 5 minutes, until you see a color change in the Benedict's reagent. You may stop testing for starch when the starch test becomes negative. To increase efficiency, leave the dropper in the test tube during this time.

4. When you can turn your attention to the next step (perhaps while you are waiting for the Benedict's reagent to react), remove the two test tubes from the 70 °C bath, pour the contents of one into the other, mixing quickly and thoroughly, and return to its bath the one tube containing the mixture. Immediately test the solution (using 4 drops, as before), for starch with the iodine solution on the spot plate. Note the color on the report sheet. (This is the "0" minute test.) Repeat every 3 minutes, as before, until the test shows no starch. Again, efficiency is improved by keeping a dropper in the test tube during this time. You also avoid contamination with another solution. You will not test this solution, or the one at 0 °C, with Benedict's reagent.

5. Repeat the routine outlined in step 4 with the two test tubes from the ice-water bath. Record the results on your Report Sheet.

52 B. THE INFLUENCE OF pH ON ENZYME ACTIVITY

NOTE: You no longer need the 70°, 0°, or boiling water bath. Only the 37° bath is needed for parts B -D.

1. Transfer 1 mL of your "stock" saliva (not the mixture prepared for Part C) to 25 mL of distilled water and mix well. Take 3 mL of this diluted saliva solution and put 1 mL in each of three test tubes. Label these tubes "5", "7", and "9." To the test tube labelled "5" add 5 mL of the pH 5 buffer provided; to the tube marked "7" add 5 mL of the pH 7 buffer, and to the tube marked tubes. Place these in the 37 °C bath. Allow all 6 test tubes to reach the temperature of the bath.

3. When the contents of the tubes are at 37 °C, pour one starch solution into one saliva solution, the second starch solution into the next saliva solution, and the last starch solution into the last saliva solution. Mix the contents of each test tube (you now should have only three test tubes containing any solution), and return the three labelled test tubes with their solutions to the water bath.

4. Immediately begin testing each of the solutions for starch, using 4 drop samples as before, with the iodine solution. Take samples from each tube every three minutes as you did for part A, until any one of the solutions no longer gives a positive test. Record your results on the Report Sheet.

52 C. EFFECT OF METAL-ION POISONS ON ENZYME ACTIVITY

Among the salts of heavy metals that are known poisons are those of mercury, lead, and copper. Their metal ions denature proteins. Enzymes are proteins.

1. Use the diluted saliva solution you set aside from part A. Transfer a 5 mL portion to each of two test tubes. To one of the test tubes add 2-3 drops of a dilute solution of the salt of a heavy metal (choose from $Cu(NO_3)_2$, $HgCl_2$, $Pb(NO_3)_2$, or whatever your instructor has provided for you.) Mix the contents thoroughly. The other test tube containing only saliva will be used as a control. Place the tubes in the 37 °C water bath (again, check the temperature.)

2. Obtain 10 mL of buffered starch (1%), the same as was used for part A. Place 5 mL portions of this in each of two test tubes, and place these in the 37 °C bath also. Allow all 4 test tubes to reach the temperature of the bath.

3. When the contents of the tubes have reached the bath temperature, pour the saliva preparation containing the heavy metal into one of the starch solutions, and the untreated saliva solution into the other starch solution. Mix each tube's contents well. You should now have two test tubes with solutions, to be returned to the water bath.

4. Immediately test each of the solutions for starch using the iodine solution on the spot plate. As before, do this every three minutes until one of them no longer tests for starch. Record your observations on the Report Sheet.

52 D. EFFECT OF THE CONCENTRATION OF AN ENZYME ON ENZYME ACTIVITY

Since the action of enzymes on substrates are those of catalysts, the concentration of the enzyme in comparison to that of the substrate is expected to be very small. However, recall the equation whereby an enzyme-substrate complex (E-S) forms before the product (P) is made:

$$E + S \rightleftharpoons ES \rightleftharpoons EP \rightleftharpoons E + P$$

Remembering that in order to have a reaction there has to be a collision, it is easy to see that the more concentrated the enzyme molecules, the greater the chance for collision, and therefore the greater the probability of reaction.

To test this, it is necessary to maintain the pH, temperature, and concentration of substrate constant in all the test solutions and to *vary only the concentration of the enzyme*. Gelatin will be the substrate, and pancreatin the enzyme, under optimum conditions of temperature, **37 °C**, and **pH**, buffered at **8**. The variations in concentration of enzyme are shown in the Report Table.

Photographic film is usually cellulose acetate covered with a uniform layer of gelatin in which some silver salt (AgBr) has been dispersed. When light hits the silver ion, it is activated, and subsequent action of the photographic developer converts it to metallic silver, so finely divided that it appears black. If this exposed and developed film is exposed to an enzyme that digests protein, the gelatin will be hydrolyzed to peptide and amino acid units which will not maintain the film structure. The silver particles will fall to the bottom of the solution, leaving the clear, colorless piece of cellulose acetate. The time required for digestion of the gelatin can be determined by the loss of black coating from the cellulose acetate.

Procedure.

1. Label 4 test tubes A, B, C, and D. Place 3 mL of the pH 8 buffer solution into each.

2. Look at the Table for Part D on the Report Sheet, and add the required amounts of pancreatin and water to each of the 4 test tubes, according to label.

3. Place all 4 tubes in a water bath at 37 °C, and allow them to reach the temperature of the bath.

4. When the test tubes are at 37 °C, drop a small square of exposed photographic film into each (be sure to use pieces that are the same size). If the film sticks to the side, push it down into the solution with a stirring rod. **Record the time.**

5. Continue to swirl the tubes in the water bath alternately, so that each gets nearly the same amount of agitation as the other.

Record on the Report Sheet the time at which all the black deposit has left the cellulose acetate film in each tube. Remove the tube from the bath, and carefully pour the solution and the film into the designated container. **Do not pour the film into the sink. It will clog the plumbing.**

Independently from your partner, plot your results on the graph provided. Use appropriate values on the axes of the graph.

EXPERIMENT 52 FACTORS AFFECTING ENZYME ACTIVITY

52 A. EFFECT OF TEMPERATURE

Time Interval (Minutes)	Color Produced by Iodine Reagent on Starch-Saliva Mixtures at			Benedict's Test
	70 °C	37 °C	0 °C	37 °C
"0" minutes				
3				
6				
9				
12				
17				
22				
27				

1. At what temperature was digestion the most rapid (optimum temperature)?

2. Explain why enzymes should be sensitive to temperature changes.

3. Are enzymes *denatured* by low temperatures?

4. What other factors influence the rates of reactions?

5. Was starch still present when reducing sugar was first detected in the hydrolysate? _____ Name the sugar:_____

52 B. EFFECT OF pH

Time Interval (Minutes)	Color Produced by Iodine Reagent on Starch-Saliva Mixtures at 37 °C and at		
	pH 5	pH 7	pH 9
"0" minutes			
3			
6			
9			
12			
17			

6. At what pH was digestion the most rapid (the optimum pH)?

Describe something other than an enzyme that will catalyze the hydrolysis of starch.

52 C. EFFECT OF HEAVY METALS

7. Write the formula of the salt of a heavy metal you were assigned:_____

Note the color of the starch-iodine mixture:

Time	"0" min	3	6	9	12	17	22
Saliva + metal							
Saliva alone							

8. Which saliva preparation, the one with or the one without the heavy-metal ion, was more effective in catalyzing the digestion of starch?

9. How might the effect of the heavy-metal ion be explained?

10. What other ions studied by your laboratory section caused inhibition of enzymatic activity (consult with your laboratory neighbors)?

11. Outline an experiment that would enable you to determine whether or not it was the negative ion from the salt of the heavy metal that adversely affected the activity of amylase in saliva. Be explicit about any assumptions the experiment might entail. (Do this on the back or on a separate sheet.)

52 D. EFFECT OF CONCENTRATION OF ENZYME ON RATE OF ENZYME ACTIVITY

Report Table:

Tube	A	B	C	D
mL pancreatin				
mL distilled water				
Starting time				
Time at end				
Digestion time				

Plot the data above as follows:

Graph of Concentration vs. time

mL of enzyme

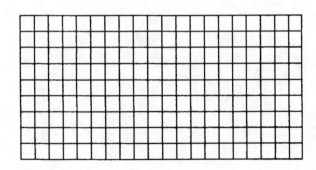

Time, in minutes

12. Draw the structure of the bond present in the peptide linkage:

13. What is the bond present in the cellulose acetate polymer? (Review Chapter 14, Carbohydrates.)

Can a peptidase hydrolyze it?_____

14. Write a short statement expressing your conclusions from the results you found for this experiment:

CHAPTER 18   Extracellular Fluids

EXPERIMENT 53 NORMAL AND PATHOLOGICAL URINE ANALYSIS

Procedure. Collect your urine sample in a clean bottle that can be stoppered. A 24-hour specimen is best but not necessary. Fill in on the Report Sheet the preliminary information before you start your tests. If possible, look at a hospital report sheet listing the substances for which urine is tested. You can identify many of these substances qualitatively, in most cases using tests you have already done for ions (Experiment 20). Use <u>clean</u> test tubes! Record your observations on the Report Sheet.

Urine Chemistry: Normal[a] Amounts Excreted in 24 Hours

Volume:	800-1800 mL	Uric acid:	250-750 mg
Creatine:	♂0-150 mg	Glucose:	0.5 g as glucose
	♀0-250 mg		0.5-1.5 g as total
Creatinine:	♂1.0-2.0 g		reducing substances
	♀0.8-1.8 g	Acetone:	none
Sodium:	40-220 mEq	Ascorbic acid:	15-50 mg
Potassium:	30-90 mEq	Bilirubin:	none
Chloride:	120-250 mEq	pH:	4.8-7.8
Calcium:	30-150 mEq	17-hydroxy steroid:	♂3-12 mg
Phosphorus:	0.9-1.3 g		♀2-10 mg
Protein:	50-100 mg	17-keto steroid:	♂8-20 mg
Amylase:	units vary with		♀5-15 mg
	method	Sulfate:	negative
Urobilinogen:	.05-3.0 mg	Lactic acid:	50-200 mg
Porphobilinogen:	0-1 mg		

[a]Values vary from laboratory to laboratory depending on the method used but are standardized in a given laboratory. These values are derived from several sources as suggestions of the "normal" range.

Chlorides. Place 5 mL of urine in a test tube, add 2 or 3 drops of dilute nitric acid and a couple of drops of silver nitrate solution. (Does the color of the urine change when it is made acid?)

Sulfates. Place 5 mL of urine in a test tube, acidify with 2 or 3 drops of dilute hydrochloric acid, heat to boiling, and add a few drops of 10% barium chloride solution. What should you see in a positive test?

Ammonium Ion. Place 5 mL of urine in a test tube, and add 2 mL of 6 M sodium hydroxide. Have a piece of moist red litmus paper ready, then carefully heat the liquid in the test tube, holding the litmus inside the test tube but not touching the walls. Do not boil the liquid. Do the vapors from the solution turn the litmus blue? Can you smell the ammonia?

Calcium Ion. To 5 mL of urine, add 5 drops of 20% potassium oxalate. A cloudy precipitate should form, gradually recrystallizing to larger crystals.

Phosphates. Acidify 5 mL of urine in a test tube with 2 or 3 drops of dilute nitric acid. Add 1 mL of ammonium molybdate solution and place the test tube in a beaker of water at 65 °C. A cloudy yellow precipitate (which usually clumps and settles) is positive for phosphate.

Now test your urine and, if possible, some abnormal urine for the following abnormal constituents: albumin, glucose, and ketone bodies. Make your report on the Report Sheet.

Albumin. Heat and Acetic Acid Test. Because albumin is a protein, it is denatured and coagulated by the action of heat. Place 10 mL of clear urine in a test tube. (If the urine is cloudy, it should be filtered.) As you hold the lower end of the tube in your hand, heat the upper portion of the urine specimen to boiling. Since phosphates, if any are present, may also precipitate by action of heat, add two to three drops of 36% acetic acid at this point to redissolve any such solids, and boil the upper portion again. A permanent, white precipitate is indicative of albumin (or other protein). Perform and compare the results of the heat and acetic acid test on normal urine and on specimens known to contain trace (0.25%), moderate (0.5%), and severe (1.0%) amounts of albumin. Attempt as best as you can to reproduce the conditions of heating uniformly for each test. Record what you see.

Glucose. Clinitest Tablets, Clinistix, or a similar product. Follow your instructor's directions. Carry out this test simultaneously on known specimens of normal urine, one with slight glucosuria (0.25% glucose), one with moderate glucosuria (0.5% glucose), and one with severe glucosuria (1% glucose). Record what you see.

[Alternate method: Benedict's Solution. A careful use of the Benedict's test makes possible a rough quantitative m estimation of the concentration of glucose in urine. Add 8 drops of filtered urine to 5 mL of Benedict's reagent in a test tube. Mix, and place the tube in a beaker of boiling water for 5 minutes. Cool the contents of the tube to room temperature. A clear and blue solution indicates the absence of glucose (or other reducing sugar). If the solution is green, the urine specimen contains roughly 0.25% glucose; if yellow, 1% glucose; if orange, more than 1% glucose; and if brick red, over 2% glucose.]

Ketone Bodies. Sodium Nitroprusside Test. This test does not differentiate between acetoacetic acid and acetone. The latter is simply a decomposition product of the former, and the two commonly occur together. Nitroprusside reagent is a finely powdered mixture of sodium nitroprusside, $Na_2Fe(NO)(CN)_5 \cdot 2H_2O$, and ammonium sulfate in a ratio of 1 g/40 g. (Test paper strips are also available commercially. Ask you instructor which procedure is to be used.)

Place 2-3 mL of urine in a test tube, add sufficient nitroprusside reagent to saturate, and shake the tube vigorously. Cautiously add 1 mL of concentrated ammonium hydroxide to form an overlayer, and note the color at the interface. A positive test is a color ranging from a faint purplish pink to dark purple. After a few minutes it acquires its maximum intensity, and then it fades to cloudy brown. Perform the test on normal urine and on urine containing acetone, and record what you see.

Name _____ Partner _____

Section _____ Date _____ Due Date _____ Score _____

EXPERIMENT 53 NORMAL AND PATHOLOGICAL URINE ANALYSIS

Urine collected between what hours_____ on _____(date)

Color_____ Odor _____ Clear/ Cloudy/ Opaque

pH (do not use litmus) _____ Method_____

For the following normal constituents place a (+) if present, a (—) if not found:

Cl^- _____ SO_4^{2-} _____ NH_4^+ _____ Ca^{2+} _____ PO_4^{3-} _____

Abnormal Constituents in Urine: Describe what you see.

Albuminuria

Normal urine_____

Slight albuminuria _____

Moderate albuminuria_____

Severe albuminuria _____

Glucosuria

Normal urine_____

Slight glucosuria_____

Moderate glucosuria _____

Severe glucosuria _____

1. List some conditions under which glucose appears in the urine.

2. Does a positive test for glucose in urine prove that the patient has diabetes mellitus?

What further test(s) are necessary, if any?

Extracellular Fluids

Ketonuria

Normal urine _____

Urine with acetone_____

Analysis of Unknowns

Obtain from your instructor two samples of pathological urine. By means of the foregoing tests, determine which, if any, of the pathological constituents are present and estimate their concentrations. Record ketone bodies as being either present or absent. With respect to glucose or albumin, however, record + + + if the condition is severe; + + if moderate; + if slight; and 0 if normal.

UNKNOWN	Albuminuria	Glucosuria	Ketonuria
Sample No. _____	_____	_____	_____
Sample No. _____	_____	_____	_____

CHAPTER 19

Molecular Basis of Energy for Living: Metabolism

EXPERIMENT 54 A MODEL OF THE CITRIC ACID CYCLE: A GROUP PROJECT

Beginning with pyruvate ion, construct ball-and stick models that show the formation of the acetyl group and its integration into the citric acid cycle as citrate ion. The formula for acetyl coenzyme A (active acetyl) can be shown by improvising a sulfur atom and attaching it to a paper marked "coenzyme A", or if some members of the group are really ambitious, the structure of coenzyme A can be made!

Start on a work bench large enough to accommodate the structures as a cycle, and as a structure is completed, put it in its proper place. Make the structures in the cycle in the salt (-ate) form, the -0⁻ of the carboxylate group can be made by putting a short stick into the oxygen ball and leaving it unfilled by a hydrogen. (A minus sign can be put on it with removable label if that helps the image.) Place arrows on paper between each structure.

If dehydration (or hydration) is involved, place a model of H_2O alongside the arrow, with a sign: (- H_2O) or (+ H_2O).

At the points where oxidation takes place, the "elements of hydrogen," H^+ and H:⁻, are formed; place an arrow to the side of the cycle leading to a hydrogen ball marked with a plus and another hydrogen ball containing a short stick and marked with a minus. Also, at the end of this arrow, place a sign "to respiratory chain" with the coenzyme involved in parenthesis. At points where the chain is shortened by removal of CO_2, place an arrow to the side of the main cycle with a molecular model of CO_2 at the end of the arrow. Make signs to identify each of the structures in the cycle. Examine the whole aggregation of models, and trace the progress of the effect of dehydration, hydration, oxidation, and decarboxylation as this remarkable coenzyme-controlled, energy-producing process moves along. Note the very high degree of oxidation of the oxaloacetate ion. In other words, there is no more energy to be derived from it until it condenses with the acetyl to become citrate. Report this experiment as directed by your instructor.

EXPERIMENT 55 A MODEL OF BETA OXIDATION: A GROUP PROJECT

Beginning with a model of a fatty acid unit such as hexanoic acid, link it with acetyl coenzyme A, and show its dehydrogenation in the beta-position forming an unsaturated acid and H^+ and $H:^-$. Add H_2O to this unsaturated acid to form a beta-hydroxy acid. This can then be oxidized to a beta-keto acid, which will react with another molecule of acetyl coenzyme A to form acetoacetyl coenzyme A and a new acyl coenzyme A containing two fewer carbons than the original palmityl coenzyme A. In all cases, show the removal of H^+ and $H:^-$, and the addition of H_2O, as well as the removal of two carbons as acetyl coenzyme A, which go to the citric acid cycle.

Continue this degradation to the point indicated by your instructor, and make your report according to the directions of your instructor.

CHAPTER 20

Chemistry of Heredity

EXPERIMENT 56 EXERCISE IN STRUCTURAL BIOCHEMISTRY: ELEMENTS OF PROTEIN SYNTHESIS

The model employed in this unit was devised by Professor Thomas Peter Bennett of Harvard University and is published by Boreal Laboratories (Tonawanda, NY) together with an accompanying booklet, *Elements of Protein Synthesis-A Guide to the Instructional Model*. By permission of both the author and the publisher this exercise is based on the Bennett model and *Guide*. Bennett's *Guide* has a thorough discussion of protein synthesis at a ribosome-*m*RNA complex.

Obtain a packet with the components of the model from the supply area. Using Fig. 40 for reference, identify the parts and sort them by type:

Ribosome	1	*t*RNA pieces	24
*m*RNA strips	4	Amino acid pieces	24

(If more than one packet is in use in the laboratory, be especially careful that parts from one are not mixed with or exchanged with parts from another.)

Select *m*RNA strip number 1 and insert it at the upper righthand corner of the ribosome into the slot made to receive it. The lettering on the strip will be upside down as seen in Fig. 31.

Amino Acid Activation. Attach amino acids to their appropriate *t*RNAs. When the pieces of each pair are joined, they represent the aminoacyl-*t*RNA system. (See Bennett's *Guide*, p. 15, top part of Fig. 18.)

Chain Initiation. In *E. coli* bacteria protein synthesis begins with the joining of one unit of formylmethionine-*t*RNA$_f$ (fMet-*t*RNA$_f$) to the first *m*RNA site at the 5′ end of the mRNA strip. The codon for fMet is AUG, the first three letters (when read rightside up) on the mRNA strip at its 5′ end. The anticodon for fMet is, therefore, UAC. Remember that cytosine (C) pairs with guanine (G) and that uracil (U) pairs with adenine (A). (The codons are written upside down on the mRNA strips because codon-anticodon pairings occur in an antiparallel direction. For the meaning of the primed numbers see page 8 of Bennett's *Guide* or your text.) Uracil replaces thymine in RNA. It is in DNA, not RNA, that thymine pairs with adenine.

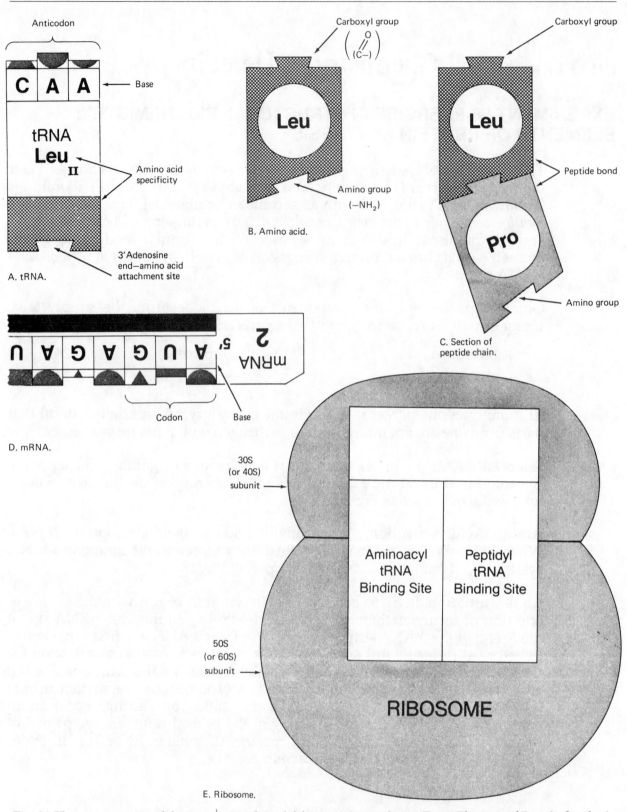

Fig. 31 The components of the instructional model for protein synthesis. From *Elements of Protein Synthesis: An Instructional Model* by Thomas Peter Bennett. Used by permission.

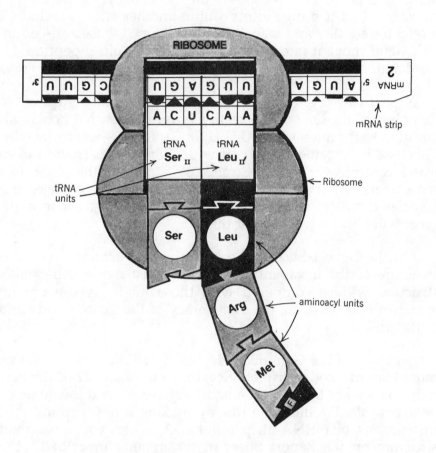

Fig. 32 Appearance of the model as a polypeptide chain grows. (From *Elements of Protein Synthesis: An Instructional Model* by Thomas Peter Bennett. Used by permission.)

Move the fitted fMet-tRNA$_f$ piece to its *m*RNA codon and over the "peptidyl-*t*RNA binding site" of the ribosome. (As a check of fitting accuracy, also be sure that the red triangles or hemispheres that form at the junctures of the pieces come out correctly.)

Chain Growth. The "aminoacyl-*t*RNA binding site" on the ribo some is now ready to receive a new aminoacyl-*t*RNA molecule. From among the models of several such molecules that you have already fitted together select the one that can be accepted by the next *m*RNA codon, which is CGU on *m*RNA strip 1. This matching brings the amino acid arginine to the mRNA-ribosome complex.

We next imagine the formation of a peptide bond. On the model simply transfer the fMet piece from its *t*RNA to the bottom of the piece labeled Arg. Remove the *t*RNA$_f$ piece just vacated by Arg and slide the *m*RNA strip to the right. As you do, carry along the second *t*RNA that now has a dipeptide hanging onto it. This reexposes the "aminoacyl *t*RNA binding site" on the ribosome, and it brings into position another *m*RNA codon CUG. Select the model for another aminoacyl-*t*RNA unit, one whose anti-codon will pair with CUG, and move it into position. Transfer the old dipeptide, fMet-Arg, to the third amino acid brought in. The development will begin to resemble Fig. 32.

Repeat the cycle of operations to illustrate further the growth of thepolypeptide. One of the relatively rare bases that occasionally appears in addition to the main four—C,G,U (or T), and A—is inosine whose symbol is I. This base is versatile. It can pair with U or C or A. Therefore, you will have to watch carefully for the appearance of "I" among the pieces of the model and remember that it can pair with any one of three bases mentioned. (The associated red-colored geometric shape hemisphere or triangle on the model piece will also vary. Watch for this.)

On the Report Sheet, following the instructions given there, record the structure of the hexapeptide you have just made using *m*RNA strip 1. Its structure will be compared with those made by the remaining pieces in exercises that illustrate the degeneracy of the genetic code and two kinds of mutations.

Degeneracy of the Genetic Code. Nearly all the amino acids can be coded by more than one codon triplet. (See, for example, p. 22 of Bennett's *Guide*.) This redundancy has the technical name *degeneracy*, and the genetic code is said to be *degenerate*. We illustrate this by making a hexapeptide under the genetic instructions of mRNA strip number 2. When you have made it, write its structure on the Report Sheet in the manner described. How are the two hexapeptides alike? Compare the two *m*RNA strips 1 and 2 with a table of codon assignments (see your text or p. 26 of Bennett's *Guide*). On the basis of these comparisons what result of using *m*RNA strips 1 and 2 *must* occur?

Mutations. Any small change in the sequence of bases on a DNA of a gene can lead to a mutation—the synthesis of an altered protein and whatever consequences that might have for the organism. Small mutations are of two types—phase-shift and base-substitution.

A phase-shift mutation is the insertion or the deletion of one nucleotide with its attached base (or of a few of such units) into or from the message. The total number of bases is thereby changed. New triplet sequences appear. The reading of the message from the point of change on will now be different. The message is faithfully read as is; there is no skipping or overlapping.

To illustrate a phase-shift mutation compare *m*RNA strips 1 and 3. Note that at position 8 of *m*RNA 3 the base A has been inserted. Make the hexapeptide according to *m*RNA 3 and compare it with the hexapeptide you made from *m*RNA 1 (the "wild type"). On the Report Sheet write the structure of the new hexapeptide.

A base-substitution mutation is the chemical change (or exchange) of one base for another. Nitrous acid, a known mutagen, can react with cytosine (C), for example, and change it to uracil (U). It can also, in effect, make adenine (A) behave like guanine (G). The total number of bases remains the same, but at the site of base substitution a different amino acid might be incorporated into the growing polypeptide. Because of the degeneracy of the genetic code, not every base substitution inevitably produces an amino acid substitution, however.

To illustrate base-substitution as a type of mutation, compare *m*RNA strip 4 with *m*RNA strip 1. The base at position 4 has been changed from C to U, and those at positions 14 and 15 from C to A. Make the hexapeptide according to the instructions of *m*RNA 4 and compare it amino acid by amino acid with the hexapeptide from *m*RNA 1. Did every mutation cause a change in the amino acid incorporated into the chain at a particular position? Write the structure of the new hexapeptide on the Report Sheet.

Name _____Partner _____

Section _____ Date _____ Due Date _____ Score _____

EXPERIMENT 56 ELEMENTS OF PROTEIN SYNTHESIS

In the table below write the structures of the hexapeptides you fashioned using the Bennett model and its four different *m*RNA strips. Let the first amino acid written be the formylmethionine (fMet) unit, and use the three-letter symbols for the amino acids that appear on their respective model parts. To illustrate this standard method of writing polypeptide structures, the pentapeptide having glycine (Gly) at the free amino end and glutamine (Gln) at the free carboxyl end with alanine (Ala), leucine (Leu), and proline (Pro), in that order, in between would be written as:

Gly-Ala-Leu-Pro-Gln

Thus the sequence for the hexapeptide made from *m*RNA strip 1 would start out as: fMet-Arg- . . . etc.

*m*RNA Used	Structure of Hexapeptide
1	
2	
3	
4	

Questions:

1. Which operation with the Bennett model illustrated that a change in the codons does not necessarily change the amino aicd inserted at that corresponding point into a growing polypeptide?

2. Which type of mutation—phase-shift or base-substitution—caused the greatest changes in the structure of the hexapeptides?

(Questions continue on next page.)

3. Correctly complete the following sentence by placing the following terms on the right lines: codon triplet, amino acid.

"Degeneracy" of the genetic code means that each _____

will be coded for by more than one_____.

4. Insulin obtained from the whale and the pig are identical in every way. Does this mean that the DNA coded for the insulin of a whale is identical with that of the pig?_____

Explain!